ANTENNA ARRAYS AND BEAM-FORMATION

Edited by **Modar Shbat**

Antenna Arrays and Beam-formation

http://dx.doi.org/10.5772/65148
Edited by Modar Shbat

Contributors

Jianwu Tao, Chang Wen-Xie, Mohamed Aymen El Cafsi, Mourad Nedil, Lotfi Osman, Ali Gharsallah, Ali Molaei, Jose Angel Martinez Lorenzo, Juan Heredia, Tiago Varum, João N. Matos, Pedro Pinho, Modar Shbat, Francisco Ordaz-Salazar, Javier S. González-Salas

CBS, Edition **2018**

Published by InTech

Janeza Trdine 9, 51000 Rijeka, Croatia

Publishing Process Manager Martina Usljebrka

Technical Editor SPi Global

Cover Designer InTech Design Team

Additional hard copies can be obtained from orders@intechopen.com

Antenna Arrays and Beam-formation, Edited by Modar Shbat
p. cm.
Print ISBN 978-953-51-3145-8
Online ISBN 978-953-51-3146-5

Contents

Preface

The main objectives behind writing this book were prompted by the recent developments in the field of wireless communication technologies, especially for the enormous wave of physical layer–related research activities. The rapid evolution of these systems could push the boundaries beyond the existed limitations and standards. Starting with the fourth-generation (4G) mobile networks and continuing to upcoming fifth-generation (5G) standards and implementations, the smart antennas with spatial processing and beam-formation capabilities are considered as completely essential approaches to be employed for forthcoming progress.

The book aims, besides collecting up-to-date contributions that are not readily available in other literature, to present and discuss the most recent research ideas in the field of antenna array design and beam-forming algorithms in a synthetic, coherent, and unified manner for the interested researchers, new comers, and experts, as well.

The chapters of the book form a grid of interlocking concepts that can be roughly categorized into the following aspects:

- The fundamentals of beam formation and direction of arrival (DOA) estimation
- Advances in antenna array design
- Advances in beam-formation algorithms

The interested reader should have a certain understanding level (solid knowledge) of many important topics such as antenna parameters and characteristics, wireless communications, estimation theory, and spatial processing schemes. Even though the propagation theory is significantly behind the presented work, no background in propagation theory is assumed. The aforementioned aspects are covered by the book chapters in quasi-sequential way, and the required topics are needed for each chapter. In general, it is highly recommended to start by reading the introductory chapter in order to obtain or recover the information needed for favorable advantages. The entire book is not dedicated to undergraduate course use; it is more for graduate-level students (without semester time frame) or researchers.

The presented research ranges from relatively straightforward mathematical analysis and derivations to simulation and empirical results. The book is designed to serve as an informative reference for the researcher involved in the analysis of the spatial signal processing techniques for smart antenna systems. Additionally, the book will help the readers (most notably wireless communication researchers) to have wider futuristic and innovative visions of the advances in the field.

The editor and the authors would like to express their gratitude to the publisher for the assigned time, invaluable experience, efforts, and staff that successfully contribute toward enriching the final overall quality of the book.

Dr. Modar Shbat
Assistant Professor in Telematics Engineering Department
Polytechnic University of San Luis Potosí, México

Chapter 1

Introductory Chapter: Smart Antennas and Beam-formation

Modar Shbat, Francisco C. Ordaz-Salazar and
Javier S. González-Salas

Additional information is available at the end of the chapter

http://dx.doi.org/10.5772/intechopen.68452

1. Introduction

Recently and over the last decade, the wireless and mobile technologies in addition to the new and improved services have grown rapidly at exponential and formidable rate. In the evolution of the modern telecommunication networks and multiple access systems, the employment of the spatial processing approaches and techniques becomes essential according to the related standards. The spatial processing is considered as the main idea behind the use of adaptive and smart antennas, antenna arrays, beamforming algorithms, interference cancelation, bandwidth-efficient signaling systems, and direction of arrival (DOA) estimation schemes (in the case of non-blind beamforming).

Smart antenna system basically consists of multiple antennas or antenna arrays and digital signal processing algorithms that are in charge of very important functions such as DOA estimation of the signals. In general, the wireless communication systems development stages can be classified based on the adopted technologies driven by the challenges of capacity demand and quality of service (QoS) requirements. These stages are summarized as follows [1]:

- Omni-directional systems: with conventional cellular structure, frequency reuse (7 cells reuse patterns), Omni-directional antenna types in the base station at the center of each cell.
- Cell splitting and sectorized systems: smaller cells (micro-cells), cell sectoring with several directional antennas in the base station.
- Smart antenna systems: with dynamic cell sectorization, multiple antennas (antenna arrays), innovative signal processing algorithms, and beamforming techniques (user location based beam assignment).

The latest telecommunication trends such as Internet-of-things (IoT) confirm the humanity willing to extend the existed technologies and employ or develop new ones which create a lot of new requirements and push the connectivity standards beyond the current limitations. In fact, some countries (like South Korea and the Netherlands) already had nation-wide IoT completed networks. Thus, the implementation feasibility (or readiness) of the IoT projects for smart homes, smart cities, and vehicles is very high and can be included in the proposed studies about smart antenna systems. For example, the mobile operator SK Telecom in South Korea installed and completed long range wide area network (LoRaWAN) based on long-term evolution (LTE) infrastructure (4G or 5G networks). This IoT network allows the smart devices from public and private sectors to receive and process data for different and various purposes. Thus, these cases can promote and trigger more efforts and investments directed to improve the wireless connectivity of the devices in such networks.

2. Smart antenna systems

The aforementioned smart antenna systems are widely implemented in two forms, namely, the switched beam approach where the system can choose one of many pre-defined antenna beam patterns (the antenna radiation or propagation pattern is defined as graphical representation of the power variation and radiation properties of the antenna as a function of the direction and space coordinates), and the adaptive array approach where the antenna adapts the radiation pattern beams in real time in accordance with the radio environment.

The smart antennas systems achieve higher capacity increase in comparison with the switched beam systems especially in the case of densely populated coverage areas and reduce more effectively the negative impacts of the interference. Additionally, there are more advantages that can be counted in favor of adaptive array systems such as range increasing, security enhancement (more difficult to tap any connection) [2], and location-based services improvements especially for emergency situations (spatial detection characteristics).

As in the case of any system or technology, some disadvantages or drawbacks of the smart antenna systems are found like the complexity of transmitters and receivers design, the high computation intensity with the need of powerful digital signal processors (DSPs), and the overall system employment cost.

At this point, two fundamental objectives should be performed by the signal processing algorithms of the smart antenna systems, namely:

- The DOA estimation for all incoming signals;
- Adaptive real-time calculation of the weights or coefficients that are used to steer and change the directions of the antenna array radiation beams toward the signal-of-interest (SOI) and at the same time to place nulls toward the signal-non-of-interest (SNOI) that is considered as interfering signal.

Hence, the smart antennas systems relay on the adaptive signal processing techniques such as DOA estimation and adaptive beamforming under the use of multiple antenna configurations (antenna arrays). Here, it is very useful to make some comments about the importance of antenna arrays in the development of the previous concepts.

http://dx.doi.org/10.5772/intechopen.68452

3. Antenna arrays

Transmit and receive diversity are effective methods for exploiting the significant benefits that are available in multiple antenna systems like multiple-input multiple-output (MIMO) wireless systems [3]. These benefits include but not limited to diversity gain (independent fading paths, channel variability reduction), array gain (average signal-to-noise ratio increase, beamforming, the gain is proportional to the array dimensions), multiplexing gain (capacity or data rate linear increase), and interference gain (aggressive frequency reuse strategy, space-time signal processing to reduce the interference effects).

It is well known that the radiation pattern generated by a single-antenna element is relatively wide with low values of directivity and gain and with less control capabilities over the important parameters. Enlarging the antenna dimensions by assembling several radiating antenna elements (array) in geometrical and electrical configurations leads to enhanced directive characteristics. The assembled antenna elements in any array can be identical (same type of antennas like dipole, micro-strip, reflector, aperture, waveguide, horn, etc.) or different. The total radiation pattern of the antenna array can be controlled and shaped using many methods such as [1]:

- The geometrical configurations (linear, planner, spherical, etc.).
- The relative distance between the elements (location and displacement).
- The amplitude and phase of the feeding electrical current for each antenna element.
- The relative radiation pattern of the individual antenna element.

The total radiation pattern of the antenna array with identical elements is obtained by pattern multiplication where the radiation pattern of a single element positioned at a reference point is multiplied by the array factor (AF). The last point can be well supported and explained by an example. Let us consider a linear antenna array with total number of identical elements equal to M with uniform spacing (d) positioned symmetrically along the same axis as shown in **Figure 1** (spherical coordinates with radial distance r, azimuth angle ϕ, and elevation or polar angle θ).

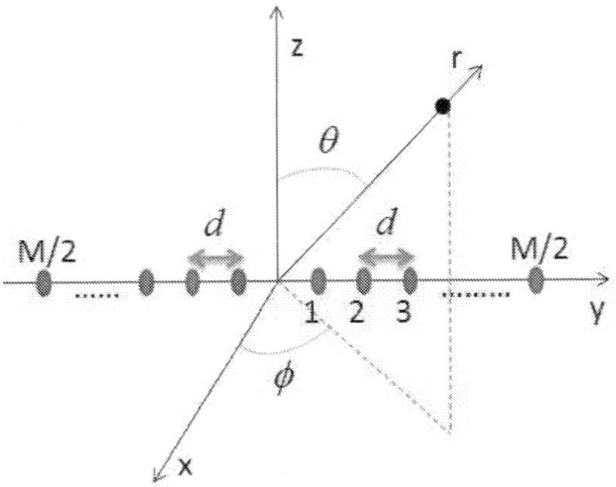

Figure 1. Linear antenna array with M identical elements.

The AF of the linear antenna array presented in **Figure 1** can be expressed using the following form [1]:

$$A\,F_M = \sum_{n=1}^{M/2} \omega_n \cos\left[(2n-1)\,\psi_n\right], \tag{1}$$

where ω_n is the amplitude of the feeding electrical current (excitation) for each antenna element, and ψ_n is given by

$$\psi_n = \frac{\pi d}{\lambda} \sin(\theta)\sin(\phi) + \beta_n, \tag{2}$$

where β_n is the phase of the feeding electrical current of the individual element, and λ is the wavelength (that shows the frequency relation with the AF definition). Thus, the total radiation pattern E_{total} presented by the amplitude of the electrical field of the linear antenna array in **Figure 1** is presented as:

$$E_{\mathrm{total}} = A\,F_M \cdot E_{se}, \tag{3}$$

where E_{se} is the single element radiation pattern located at the array reference point. One important observation form the last discussion is that by changing the values of the AF coefficients ω_n and β_n, it is possible to control the shape of the radiation pattern plus the major to minor lobes level and the scanning capabilities of the antenna array, respectively. Obviously, any beamforming technique is able to use the previous control coefficients to shape and redirect the radiation lobes or beams in accordance with the user location. In the case of mobile communication, the planner arrays are preferred according to the three dimensions (3D space) scanning abilities.

4. Antenna beam-formation

The beam formation (BF) is a spatial signal processing technique coupled with multiple antennas (antenna array elements) that are adaptively phased to form, direct, and concentrate the beams of the radiation pattern [4]. The BF algorithms can be implemented at both transmitter side (transmit beamforming) and receiver side to provide significantly improved array gain, higher signal-to-noise ratio (SNR), and considerable reduction in co-channel interference owing to the spatial selectivity of the directional antenna array elements.

For the millimeter wave (mmWave) systems, almost all modern wireless communication networks, the antenna BF shows great benefits since highly directional adaptive antenna array elements can be designed with low profile and steering capabilities in various directions to meet and coherently align the SOIs and dampen the undesired or interfering signals (SNOIs). In **Figure 2**, a normalized radiation pattern of linear antenna array with eight identical elements ($M = 8$) and equal spacing is presented. It is shown that by employing BF technique, the major beam (main lobe) is directed toward the SOI ($\theta_1 = 30°$), and a null is placed toward the SNOI ($\theta_2 \approx 50°$).

The next figure (**Figure 3**) presents a simple adaptive BF block diagram for two-element array with spacing $d = \lambda/2$ receiving the desired signal SOI at θ_1 and the interfering signal SNOI at θ_2. This example illustrates the basic concept of null formation to satisfy certain radiation pattern requirements by BF weights (w_1, w_2) computing.

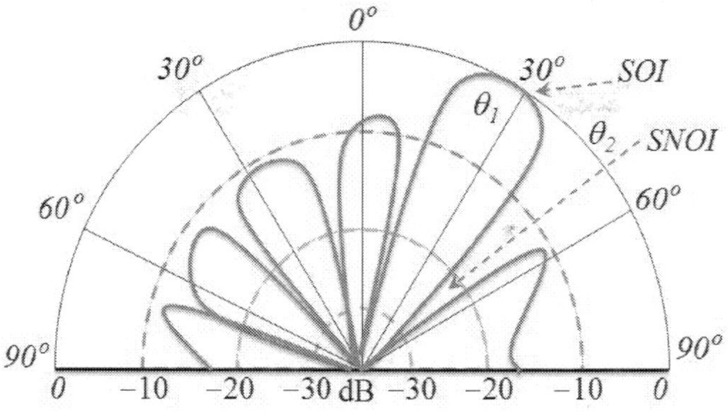

Figure 2. Antenna radiation pattern of linear antenna array with $M = 8$ identical elements and BF technique.

The BF technique in **Figure 3** should determine the complex weights w_1, w_2 to receive the desired signal $S(t)$ and cancel the interfering signal $I(t)$. Thus, the array output $y(t)$ due to the beamforming process will ideally contain the desired signal only and totally reject the interference. Solving two complex equations, the optimum complex weights w_1, w_2 are defined to achieve the maximum signal-to-interference ratio (SIR). In practice, and under non-stationary signal condition and non-homogeneous noise, the BF complex weights are computed with adaptive algorithms.

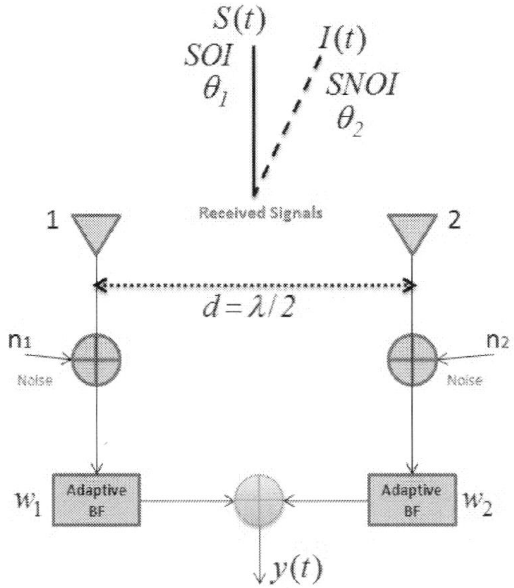

Figure 3. Antenna BF block diagram.

Some kind of adaptive beamforming algorithms do not need the information supplied by the DOA estimation as in the case of DOA-based adaptive beamforming algorithms (non-blind beamforming). Instead, these algorithms (blind) uses reference signals or training sequences (codes) in order to adjust the amplitude and phase coefficients (ω_n, β_n) of the antenna array factor (AF) [5]. A common set of predefined BF codes can be found in the related standard (IEEE 802.15.3c).

In wireless communication sector, the adaptive BF weights are chosen to maximize the quality of communication channel (or the quality of the received signal). Some commonly used adaptive BF approaches can be mentioned:

- Minimum mean-square-error (MMSE) approach: the complex weights are defined in order to minimize the mean square error between the beamformer output and the expected signal (using Wiener filter) [6].
- Least mean square (LMS) approach: very simple and effective algorithm that minimizes the mean squared error (MSE) cost function and computes the BF weights using iterative and bounded conditions.
- Maximum SIR approach: the complex weights are determine to maximize the SIR value (desired signal and interference strengths are estimated by the receiver).
- Minimum variance approach: the complex weights are computed to minimize the noise variance at the beamformer output (the signal waveform and DOA are known).

The beam-space adaptive beamforming (beam-space transformation) employed in communication systems uses fast Fourier transform (FFT) beamforming where a set of FFT outputs can be combined using complex weights and sums to form arbitrary radiation patterns (baseband signals are combined from different antenna elements) [7].

5. DOA estimation

In some other references, the DOA estimation is called angle of arrival (AoA) or angle of departure (AoD) estimation. The ability to measure the DOA of a wireless signal with higher resolution in comparison with the antenna beam width is defined as super resolution. The type of beamformers that needs DOA estimation is called non-blind BF techniques. When the antenna array elements receive the incoming signals from all directions, the DOA technique estimates these directions based on the time delay and array geometry concepts. To understand these concepts, the two elements antenna array in **Figure 4** can help to simplify and derive the DOA estimation.

The arrival time difference Δt of the signal at the two antenna elements (assuming that the signal is direct from the source or we have a plan wave) can be given by

$$\Delta t = t_2 - t_1 = \frac{d\cos\theta}{c} \tag{4}$$

where c is the speed of light in the free space. Clearly, the DOA demonstrated by the angle θ can be determined knowing the spacing d between the array elements (by design) and the time delay Δt. This method is very sensitive to signal reflections (multipath problem) and to the existence of interfering signals.

The DOA estimation algorithms are classified based on the data analysis into four different groups:

- Conventional algorithms: the estimation process is based on beamforming and null steering without the exploiting of the received signal statistics such as delay-and-sum method and Capon's minimum variance method.
- Sub-space based algorithms: the method utilizes the received signal structure to improve the resolution such as multiple signal classification (MUSIC) and the estimation of signal parameters via rational invariance technique (ESPRIT) [1].
- Maximum likelihood (ML) algorithms: the estimation is obviously optimal based on the maximum likelihood sense where the DOA algorithm maximizes the likelihood that the signal arrived from a particular direction [8]. The ML algorithm outperforms the sub-space–based techniques under low SNR and spatially correlated signal conditions, but it is computationally intensive.
- Integrated algorithms: the property restoral method and the sub-space–based approach are combined in order to separate multiple signals and estimate their spatial signatures prior to the DOA estimation (mainly performed by sub-space based algorithm).

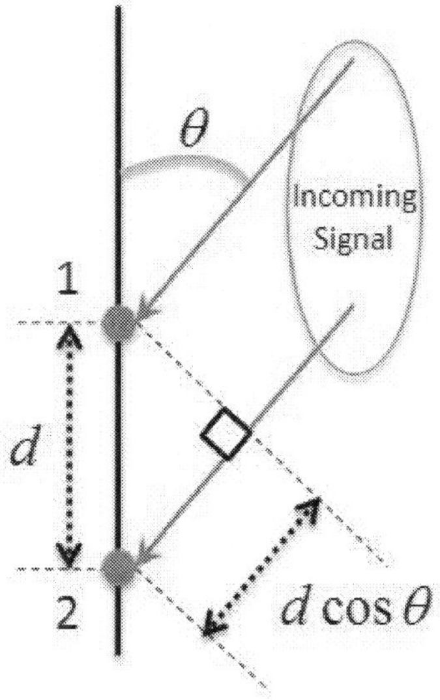

Figure 4. DOA estimation main concept.

Since the DOA is a parameter estimated from received data, the Cramer-Rao lower bound (CRLB) can be used to define the minimum variance of this estimation [9]. The CRLB defines the best performance obtained after minimizing the residual noise in unbiased estimates (noisy data defiantly produce noisy estimates). Thus, the determination of the CRLB for any DOA algorithm is helpful to find the theoretical limits for performance evaluation.

6. Beam-formation performance remarks

Two-dimension linear antenna array pattern with 10 elements is demonstrated in **Figure 5** in the case of LMS beamforming algorithm. In this scenario, the antenna array receives two SOIs at $\theta_1 = 0°; 10°$ and two SNOIs at $\theta_2 = 30°; -40°$ where the SNR = 5[dB] and the interference-to-noise ratio is equal to INR = 10[dB] (the interference is more severe problem comparing with the noise).

It is noticeable that the radiation pattern generated by the antenna array elements and LMS algorithm has its maximum toward the SOIs directions while the nulls (minimum power) are placed toward SNOIs directions.

The time-varying multipath channel (fading channel) is a common model used for wireless communication networks. The spatial diversity techniques (replicate the transmitted information over independent channels using different antennas) form an effective solution to combat the negative influence, presented by performance degradation, of the fading channels. Any BF approach can be combined with spatial diversity to improve and enhance the performance of the wireless transmission. The performance simulations given in **Figure 6** are made in the case of Rayleigh fading channel model and for transmit diversity scheme (two antennas) with and without BF. The bit error rate (BER) as a function of energy per bit E_b to noise spectral density N_0 ratio (similar to SNR) is presented for binary phase shift keying (BPSK) modulation with hard decision receiver.

The results shown in **Figure 6** indicate that the BER is less using BF (LMS algorithm) in comparison with the case when the BF is not employed over the whole E_b/N_0 range. Thus, the BF technique could boost and improve the receiver performance by reducing the BER value.

The appropriate utilization of the channel state information (CSI) sent over the feedback link to the transmitter side helps to achieve the highest performance from any MIMO or antenna array system. The beam training (alternative solution of the conventional feedback) is employed to configure transmit and receive BF vectors where the transmitter sends information over several beams and uses the feedback from the receiver to find the beast beam (IEEE 802.11ad) [4]. New approaches have been developed for closed-loop spatial multiplexing based on the beam training and feedback combination such as code-book BF, BF with weight optimization based on AoD estimation, and long-term BF [4].

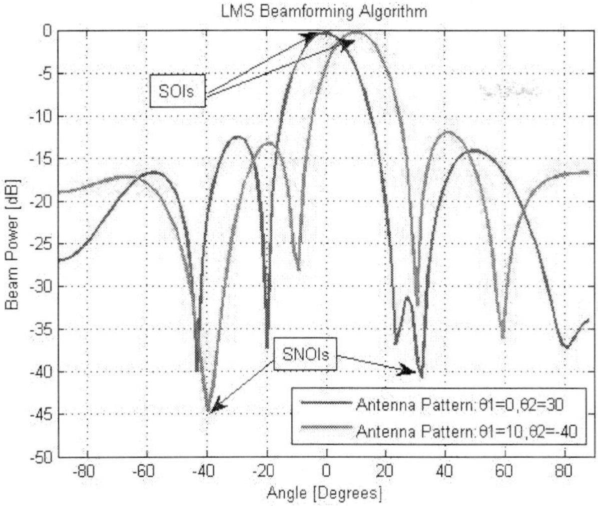

Figure 5. 2D antenna array pattern of 10 elements applying LMS-BF algorithm.

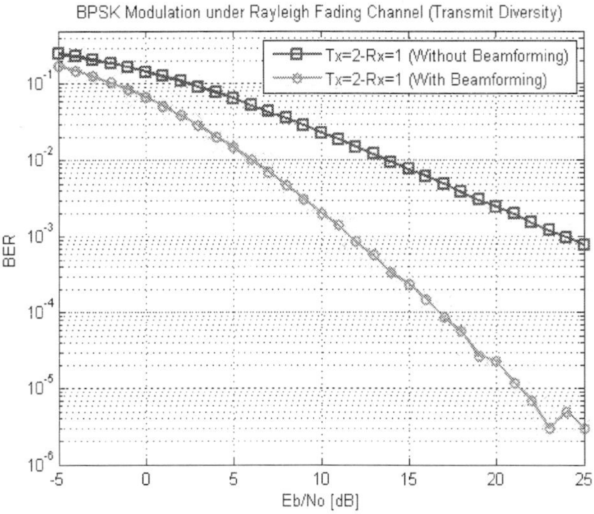

Figure 6. BER over Rayleigh fading channel with and without BF.

The basic concepts and the importance of smart antennas, BF, and DOA estimation are addressed in this chapter. The book focuses on the latest contributions made by researches and experts on smart antennas and beam-formation approaches in order to present the recent advances on the field.

Acknowledgements

This work is performed by Radio Physics Research Group (RPRG) at Polytechnic University of San Luis Potosi (UPSLP) and supported by the National Council of Science and Technology in Mexico (Consejo Nacional de Ciencia y Tecnología CONACYT, Mexico).

Author details

Modar Shbat*, Francisco C. Ordaz-Salazar and Javier S. González-Salas

*Address all correspondence to: modar.shbat@upslp.edu.mx

Telematics Engineering Department, Polytechnic University of San Luis Potosí, San Luis Potosí, SLP, México

References

[1] Balanis CA. "Antenna Theory: Analysis and Design". 3rd ed: A John Wiley & Sons, INC. Hoboken, New Jersey; 2005

[2] Yang N, Wang L, Geraci G, Elkashlan M, Yuan J, Di Renzo M. "Safeguarding 5G wireless communication networks using physical layer security". IEEE Communications Magazine. April 2015, Vol. 53(4): pp. 20-27

[3] Proakis JG, Salehi M, "Digital Communications". 5th ed: McGraw-Hill Higher Education, New York; 2008

[4] Sun S, Rappaport TS, Heath RW, Jr., Nix A, Rangan S. "MIMO for millimeter-wave wireless communications: Beamforming, spatial multiplexing, or both?". IEEE Communications Magazine. December 2014, Vol. 52(12): pp. 110-121

[5] Love DJ, Heath RW, Strohmer T. "Grassmannian beamforming for multiple-input multiple-output wireless systems". IEEE Transactions on Information Theory. Oct. 2003; Vol. **49**(10): pp. 2735-2747

[6] Mallaparapu U, Nalini K, Ganesh P, Raghavendra T, Ullan Khan H, Prasanna DL, Madhav BTP. "Non-blind adaptive beamforming algorithms for smart antennas". IJRRAS. 2011; Vol. **6**(4): pp. 491-496

[7] Vorobyov SA, "Adaptive and Robust Beamforming". in Academic Press Library in Signal Processing; Vol. 3. Array and Statistical Signal Processing: Academic Press. Elsevier Ltd, Amsterdam, Netherland; 2014: pp. 503-552

[8] Vorobyov SA, Gershman AB, Wong KM". Maximum likelihood direction-of-arrival estimation in unknown noise fields using sparse sensor arrays". IEEE Transactions on Signal Processing. Jan. 2005; Vol. **53**(1): pp. 34-43

[9] Kay SM, "Fundamentals of Statistical Signal Processing: Estimation Theory". Vol. **1**. New Jersey, USA: Prentice Hall; 1993

Chapter 2

The Design of a 360°-Switched-Beam-Base Station Antenna

Mohamed Aymen El Cafsi, Mourad Nedil,
Lotfi Osman and Ali Gharsallah

Additional information is available at the end of the chapter

http://dx.doi.org/10.5772/67522

Abstract

The concept of switched-beam antenna (SBA) systems covering an area of 360° for wireless base station applications is presented. First, a reconfigurable pattern antenna (RPA), which is composed of an omnidirectional slot-antenna array surrounded by an active cylindrical frequency selective surface (FSS), is studied. The behavior of FSS is controlled by PIN diodes which are able to divide the Azimuth plane into six sectors from one common source. Unfortunately for a sector antenna, a huge number of diodes are required which complicates the structure in terms of efficiency and complexity. However, a simple and an efficient SBA configuration based on a hexagonal Fabry-Pérot cavity leaky wave antenna (FPC LWA) arrays is proposed as a solution for RPA problems. A sector-directive beam is generated from a simple patch antenna embedded inside a resonant Fabry-Pérot cavity with specific dimensions which have an influence on beamwidth and radiation efficiency. To increase more sectorization level and channel's capacity, the proposed sector in FPC LWA arrays can be divided into three subsectors by using an active high-impedance surface (HIS). As a conclusion, SBA based on FPC LWA is the most suitable solution for future wireless communications.

Keywords: switched-beam antenna, base station, reconfigurable pattern antenna, frequency-selective surface, Fabry-Pérot cavity leaky wave antenna

1. Introduction

Wireless communications are evolving to offer new generations of services such as video, satellite broadcasting, Internet, and mobile networks, which require high throughput to transmit multimedia signals with a good reliability [1]. In fact, these requirements force the improvement of the immunity of base station in front of many constraints such as multipath fading and interferences created between wireless networks due to an omnidirectional radiation pattern [2].

Beamforming is seen as a solution for the problems cited previously to ensure a high quality of service (QoS) and increased channel capacity using smart antenna systems [3].

Basically, a smart antenna system is based on the theory of phased antenna arrays wherein the differential phase between adjacent antennas permits the scan of radiation pattern. Depending on their architectures, these systems are divided into two categories based on transmit strategy: adaptive antennas and switched-beam antenna (SBA) [4].

Adaptive antennas are able to adapt their radiation patterns by exploiting the wireless radio channel at both transmitter and receiver sides with high precision using digital signal processing. Moreover, these adaptive systems have the ability to distinguish between user signal and interferences and also to steer the main beam in the desired direction with insertion zeros, thereby increasing the signal-to-interferences ratio [2].

However, the physical implementation of this system is very complex and expensive and requires high-power consumption and a longer time period to compute the current weight's values (amplitude and phase) for each antenna which is not suitable for high data rate communications [4].

On the other side, the SBA generates multiple fixed beams in a specific area. The behavior of these systems consists in the detection of the user's position by measuring signal strength, selects one of the predefined fixed beams and switches from one beam to another when the user moves [4].

In general, switching operation is assured by a fixed beamforming network, such as Blass matrix [5], Rotman lens [6, 7], and Butler matrix [4], connected to a linear antenna array. When interference and useful signal are both located in the same beam, the SBA system is disabled to resolve this scenario compared to the adaptive antennas [8]. Thus, the SBA is less costly, but its complexity and losses increase when the number of beams is increased.

In addition, these two classical smart antenna systems do not provide 360° of coverage due to the printed circuit broad (PCB) technique used on the design of antenna which radiates above the ground plane of the structure allowing only half-plane coverage. Moreover, if the antenna element has an omnidirectional radiation pattern, such as monopole, the directivity will be divided into two opposed directions that are decreasing, which causes the radiated energy.

Two SBA configurations, allowing 360° coverage, are presented in this chapter, which is organized as follows. After the introduction, an RPA is presented using as a technique to create SBA, which is obtained from an omnidirectional source surrounded by an active FSS, wherein active elements are mounted on its surface. However, this type of configuration requires a huge number of switchers such as PIN diodes that complicate the structure. To overcome the use of the PIN diodes and achieve directive beam from one low-gain source, Section 3 presents a simple and an efficient hexagonal SBA configuration based on Fabry-Pérot cavity leaky wave antenna (FPC LWA) arrays. Finally, the last section concludes the chapter.

2. Reconfigurable switched-beam antenna using an active cylindrical frequency-selective surface

2.1. FSS unit cell

FSS is an artificial electromagnetic band gap (EBG) metamaterial periodic structures which has the ability to control the propagation of electromagnetic waves, as spatial filter, in a specific frequency band [4, 9].

This behavior can be used as a beamforming technique for covering an area of 360° and creating a smart antenna. The proposed periodic unit-cell structure presented in **Figure 1** is composed of a discontinuous metal strip printed on a flexible substrate where a PIN diode is inserted in the discontinuity [9].

According to the PIN diode state, the strip has a reconfigurable transmission/reflection coefficient [10]. When the diode is turned ON, it is equivalent to a resistor and acts as a reflector. While for OFF state, the diode is equivalent to parallel RC circuit allowing the strip to be transparent to the electric field of a transverse electric (TE) plane wave which is parallel to it [9, 10].

Figure 2 illustrates the reflection/transmission coefficients of the proposed FSS unit cell.

A high reflection coefficient R is obtained for the ON state, whereas for the OFF state, a high transmission coefficient T is achieved.

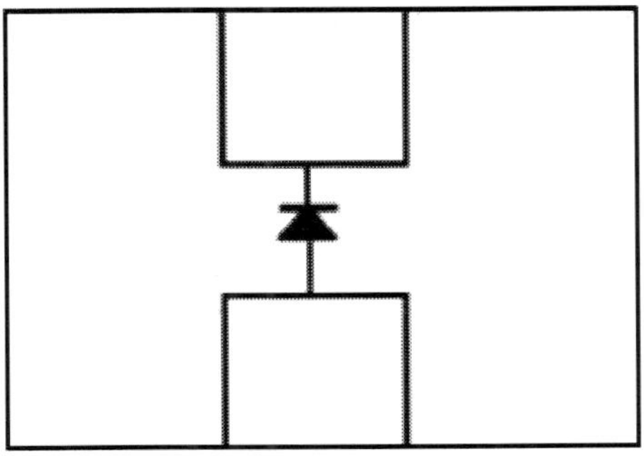

Figure 1. FSS unit cell.

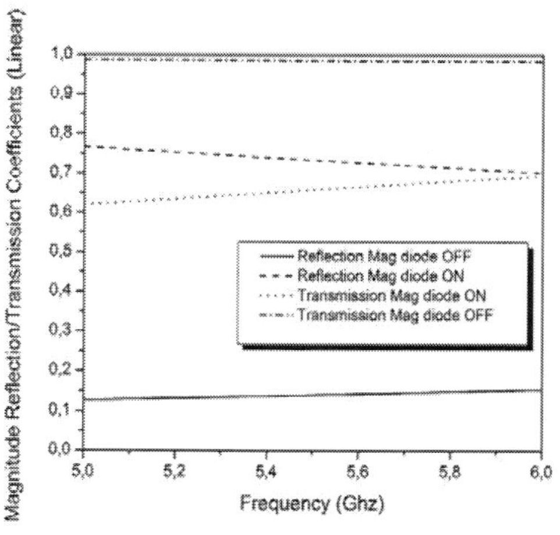

Figure 2. Magnitude of reflection/transmission coefficient of FSS unit cell.

Indeed, R and T are inversely proportional to the ON-state resistance and the OFF-state capacitance values, respectively [10].

So, the diode should have a low resistance's value in ON state and a low capacitance's value in OFF state. For the proposed design, the PIN diode GMP-4201 from MicroSemi is used and modeled with the ON-state resistor of $R_{ON} = 2.3\ \Omega$ and, in the OFF-state, with parallel RC circuit with $R_{OFF} = 30\ \mathrm{K}\Omega$ and $C_{OFF} = 0.18$ pF [10]. The surface of the gap should be equal to the diode's dimensions.

2.2. Reconfigurable pattern antenna

In fact, RPA is achieved by placing an omnidirectional antenna in the center of the active cylindrical FSS. A sector notion is defined in this design by dividing the cylinder into two semi-cylinders. The first semi-cylinder is activated by turning its PIN diodes "ON" and behaves as a parabolic reflector which defines the directivity of radiation pattern. However, the other diodes are turned OFF and the second semicylinder will be transparent to the electromagnetic waves allowing radiation [10].

So, by controlling PIN diode states, a dynamic FPC antenna is obtained which is able to scan all azimuth planes from an omnidirectional source.

2.2.1. Omnidirectional slot-array antenna

If the source used is a simple dipole, the system will suffer from a low directivity due to the small aperture surface of FSS [11]. To enhance the directivity, authors in [10] have used a dipole array wherein pipes are used between two successive dipoles to ensure impedance matching and in-phase excitation of the array. Again, the use of pipes complicates the system in terms of adaptation.

To overcome problems cited above, a high-gain omnidirectional radiation pattern can be obtained from a printed slot array which is fed by a strip line embedded in sandwich between two substrates with the same thickness as shown in **Figure 3**. The radiation phenomenon is done from the two ground planes situated at the top and bottom of the structure [9].

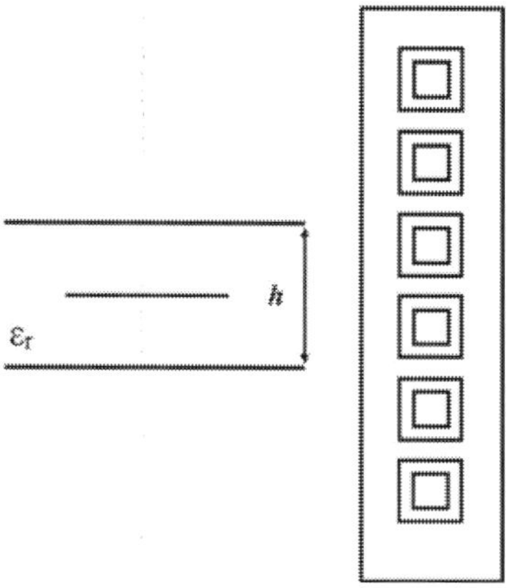

Figure 3. Omnidirectional slot-antenna array.

Return loss and radiation pattern in H- and E-planes are shown in **Figures 4** and **5**, respectively. The structure operates at 5.8-GHz central frequency of Industrial Scientific Medical (ISM) band with 1.72% of the bandwidth.

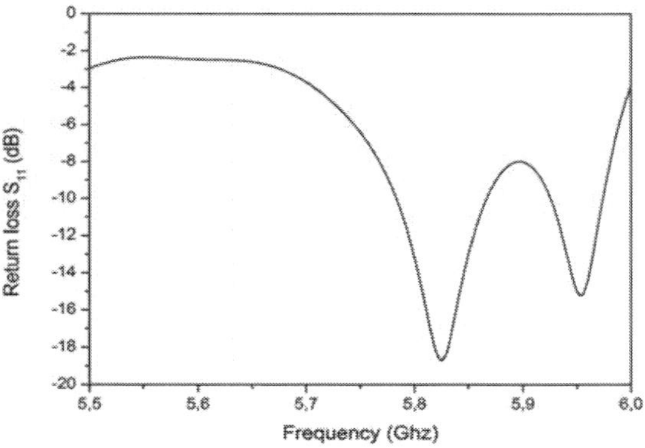

Figure 4. Return loss of slot-antenna array.

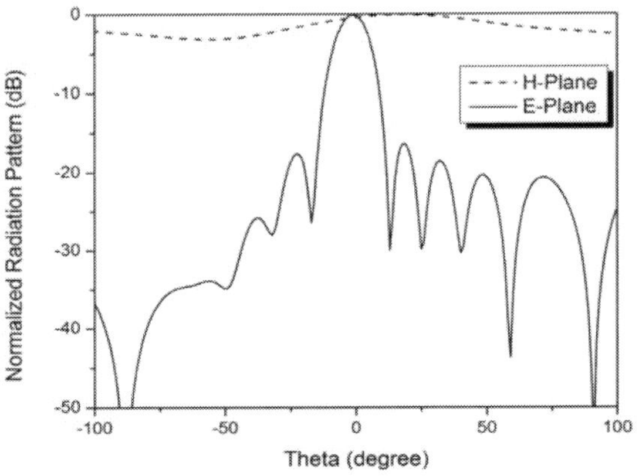

Figure 5. Radiation pattern in H- and E-planes.

Due to its constant values in all directions and according to **Figure 5**, the reconfigurable radiation pattern is located in H-plane and will be used to divide the azimuth plane in many fixed sectors.

2.2.2. Switched-beam antenna system

The whole reconfigurable antenna system, which is composed of the slot-antenna arrays surrounded by the active conformal FSS, is given in **Figure 6**.

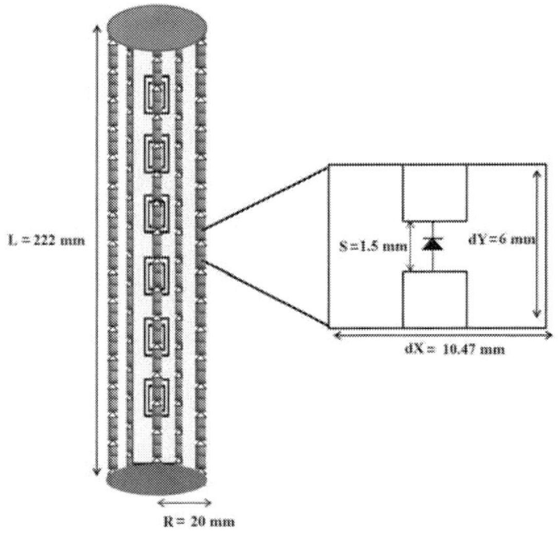

Figure 6. Reconfigurable pattern antenna.

http://dx.doi.org/10.5772/67522

A recent study reported in [12] demonstrates that six-sector deployments increase the channel capacity by 70–80% compared to traditional three sector sites. Therefore, FSS parameters, which are the number of strips N, the width of unit cell D_x, and the radius R, are studied to satisfy this requirement using the following equations [10]:

$$N = \frac{360}{P_\theta} \tag{1}$$

$$D_x = R \,.\, P_\theta \cdot \frac{\pi}{180} \tag{2}$$

where P_θ is the period angular and, in this case, is equal to 30°.

The activated strip number, whose PIN diodes are turned ON, has an influence on the H-plane beamwidth and equals 4, in the proposed design, to achieve a symmetrical half beamwidth of 70° as shown in **Figure 7**.

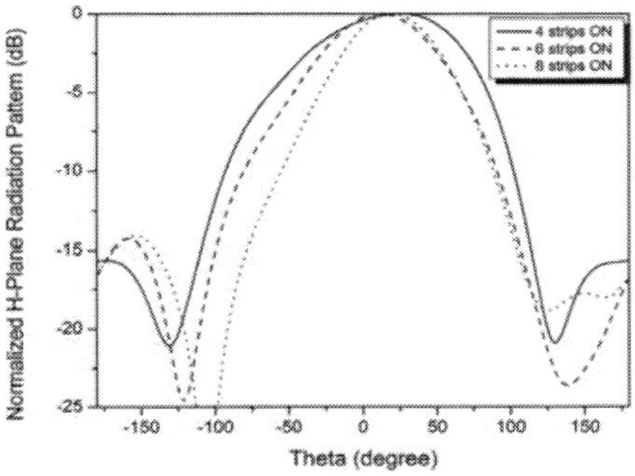

Figure 7. Effect of strips in ON state on H-plane radiation pattern.

The effect of FSS radius on radiation pattern is presented in **Figure 8** where the optimum value found is close to the half wavelength of the operating frequency in free space.

Figure 9 illustrates six switched beams covering all azimuth planes.

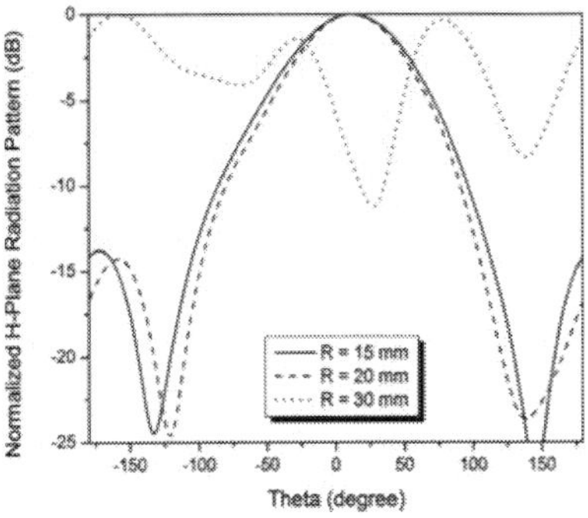

Figure 8. Effect of FSS cylinder radius on H-plane radiation pattern.

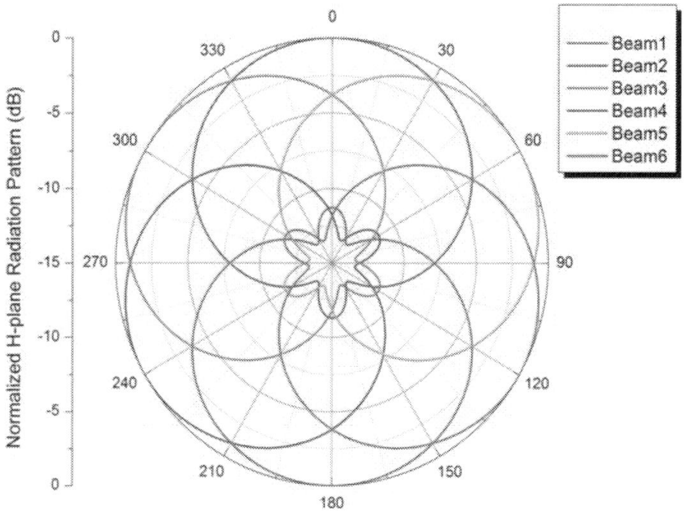

Figure 9. Switched-beam antenna.

With these optimum parameters of FSS, a directional sector antenna is achieved offering 12 dBi of directivity with a wide beam in H-plane and a narrow beam in E-plane. The proposed antenna is surrounded by 12 strips along the antenna length and each strip requires 36 PIN diodes. As a result, a huge number of active elements are required, which decreases the radiation efficiency and increases the cost and complexity of the system.

3. An efficient hexagonal switched-beam antenna structure based on Fabry-Pérot cavity leaky wave antenna

SBA system based on a planar FPC LWA is proposed, in this section, as a solution to avoid the use of PIN diode on an RPA system.

In general, a planar resonant FPC is composed from an x-directed horizontal electric dipole (HED) backed with a ground plane and PRS superstrate placed on top of the structure [13, 14] as shown in **Figure 10**.

This type of structure is seen as a leaky parallel waveguide or two-dimensional (2D) leaky wave antenna which is able to produce a pencil beam at broadside under a specific condition [14, 15]

$$\frac{h}{\lambda_0} = \frac{0.5}{\sqrt{\varepsilon_r - \sin^2(\theta_i)}} = \frac{0.5}{\cos(\theta_i)} \qquad (3)$$

where h is the distance between the ground plane and PRS screen, λ_0 is the free space wavelength, ε_r permittivity of the medium of propagation, and θ_i is the incident wave angle.

http://dx.doi.org/10.5772/67522

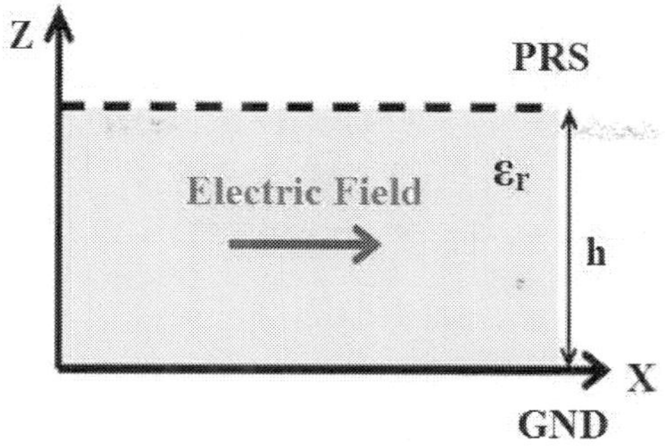

Figure 10. Two-dimensional leaky wave antenna.

For air-filled cavity (ε_r = 1) and at broadside (θ_i = 0°), the incident wave propagation has a normal angle to structure providing a distance h equal to half wavelength of the operating frequency in free space.

In this proposed design, a simple patch antenna, printed on the Rogers RT5880 substrate (permittivity ε_r = 2.2 and thickness h = 1.575 mm), is used as HED and resonates at the frequency 5.8 GHz. However, PRS is made from an inductive metal strip grating (MSG) [16] screen printed on Rogers RT6010 superstrate (permittivity ε_r =10.2 and thickness t = 2.5 mm) with the following parameters: width of strips w = 5 mm and period of strips P = 7 mm) [4, 17].

The proposed FPC LWA is presented in **Figure 11**.

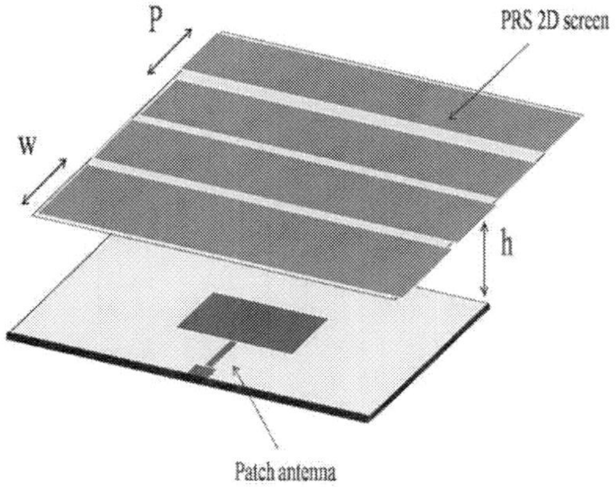

Figure 11. Fabry-Pérot cavity leaky wave antenna.

3.1. Azimuth division approach

Our sectorization approach consists of dividing the azimuth plane in N sectors where N is defined according to an angle ϕ formed between two successive cavities FPC LWA forming a specific shape [4, 17].

By fixing the size on y-direction, a superstrate portion ΔL is added in two edges to interconnect the two adjacent PRS screens and, in this case, we talk about one-dimension (1D) FPC LWA which enhances the gain of the patch antenna by a factor about 3 dB [4].

Furthermore, PRS length controls the attenuation constant α of the propagation constant k inside cavity which has an influence on the radiation efficiency [18, 19].

The leakage rate or attenuation coefficient α and radiation efficiency (η_{RA}) are obtained using the following formulas, respectively [20]:

$$\alpha = \frac{S_{SLL}}{8.686L} \tag{4}$$

$$\eta_{RA}(\%) = 1 - \exp\left[-2\alpha\left(\frac{L}{2} + \Delta L\right)\right] \tag{5}$$

where S_{SLL} is the side-lobe level (in dB) and L is the total length of PRS along x-direction.

The effect of sector number on antenna radiation efficiency is summarized in **Table 1** and **Figure 12**.

Angle (in degree)	ΔL (in mm)	Attenuation coefficient α (in m^{-1})	Directivity (in dBi)	Number of sectors	Side-lobe levels (in dB)	Total length of PRS (in mm)	Radiation efficiency (%)
10	2.5	0.021	10.52	36	−12.2	65	74.46
20	5	0.02	10.5	18	−12.2	70	75.34
30	7.5	0.018	10.53	12	−12.2	75	74.07
40	10	0.0175	10.74	9	−12.5	80	75.34
60	16	0.0211	11.39	6	−16.2	92	85.64
90	28	0.013	10.55	4	−13.3	116	77.86
120	46	0.0123	10.3	3	−16.2	152	84.585

Table 1. The effect of sector number on antenna radiation efficiency.

The optimal radiation efficiency is achieved for six and three sectors and its values are close to the optimum radiation efficiency (87%) predicted in [20].

However, a directivity of 11.39 dBi is achieved for six sectors compared with 10.3 dBi for three sectors despite the large length of PRS [4]. This observation can be explained by the existence of an optimal size for the cavity which limits radiation efficiency.

The whole SBA FPC-LWA system is illustrated in **Figure 13**. It is composed of the hexagonal FPC LWA arrays connected to a Single Pole 6 Throws (SP6T) RF switcher [21] which switches the common input power between six outputs.

http://dx.doi.org/10.5772/67522

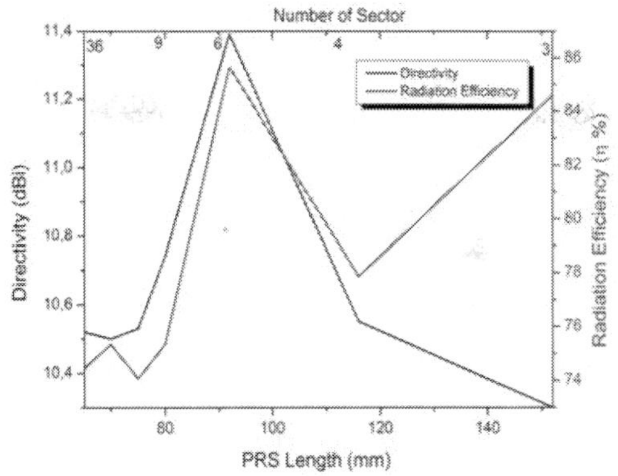

Figure 12. Radiation efficiency and directivity versus PRS length and number of sectors.

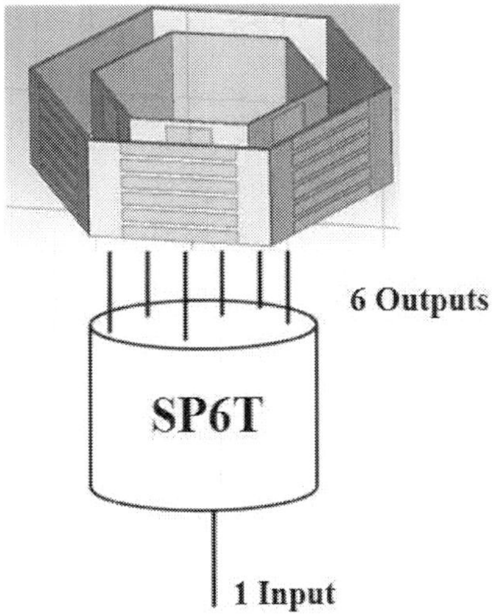

Figure 13. The proposed hexagonal switched-beam antenna system.

The photograph of fabricated prototype is shown in **Figure 14**.

The simulation and experimental results of return loss are shown in **Figure 15**. From these curves, it can be concluded that a bandwidth of 100 MHz is obtained for a return loss below 10 dB. Due to the repetitive recurrence of the structure, all coverage directions have the same bandwidth which represents 1.46% of the total bandwidth at the central frequency 5.8 GHz of ISM band.

It can be seen also that measured result is shifted on the left from the simulation caused by the height of cavity which exceeds slightly a half wavelength found by Eq. (3).

Figure 14. Fabricated prototype.

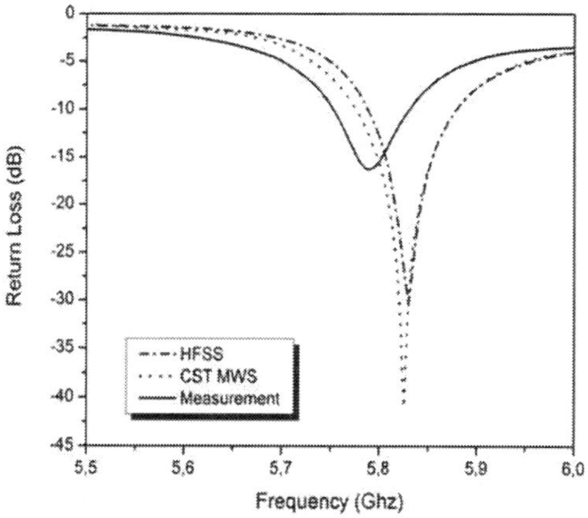

Figure 15. Return loss.

Simulated and experimental radiation patterns in the azimuth plane or E-plane are shown in **Figure 16** wherein a good agreement is carried out. A sector is defined by 3-dB beam width around 30°,which is the half value found in RPA [4], and according to [22], this beamwidth value allows to decrease the overlap between beams to 9 dB compared to 3 dB (the case of beamwidth of 60°) increasing thus signal-to-noise ratio (SNR).

The activate radiation pattern has a back-lobe level and a side lobe level of 15 dB. Some beams, in measurement, have high SLL which can be explained by the imperfect interconnection between two successive FPC LWA introducing parasite radiations.

http://dx.doi.org/10.5772/67522

It is important to see how the six beams cover all azimuth planes without any crossover between adjacent beams.

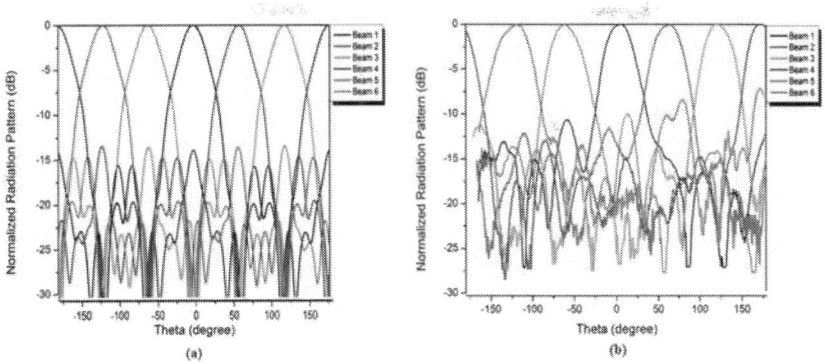

Figure 16. E-plane radiation pattern: (a) simulated and (b) measured.

Figure 17 presents the results of gain in the azimuth plane between 5.6 and 6 GHz found by measurement and simulation using two full-wave analyses Ansoft HFSS [23] and CST Microwave studio [24].

We noted an intersection between the measured value and the simulated one found by HFSS. This observation confirms our previous study which proposed that FPC LWA has a high radiation efficiency as presented in Section 3.1.

As result, a gain around 10 dBi is obtained for both simulation and measurement. However, a simulated value of CST Microwave studio is greater than HFSS by 1 dB, which can be explained by the different methods used in the simulation process.

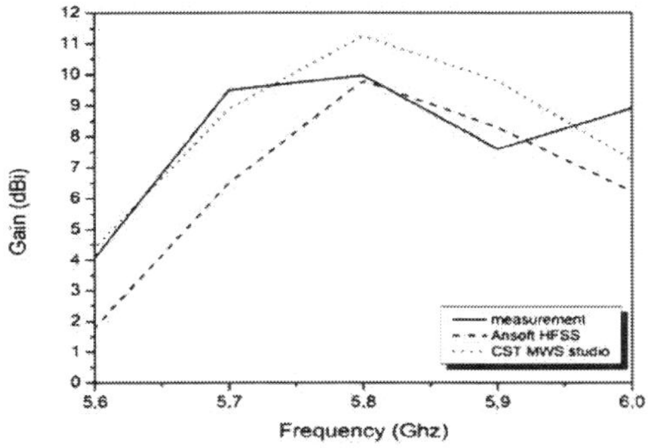

Figure 17. Simulated and measured gain.

3.1.1. Sector division

Dividing the sector into subsectors or small cells further enhances the channel's capacity by increasing the sectorization level. As an example, an active high-impedance surface (HIS) based on electromagnetic band gap mushroom-like structure, located on the ground plane of 50 ohms fed coaxial patch antenna [3], can be used as beamforming technique to scan the beam inside sector as shown in **Figure 18**.

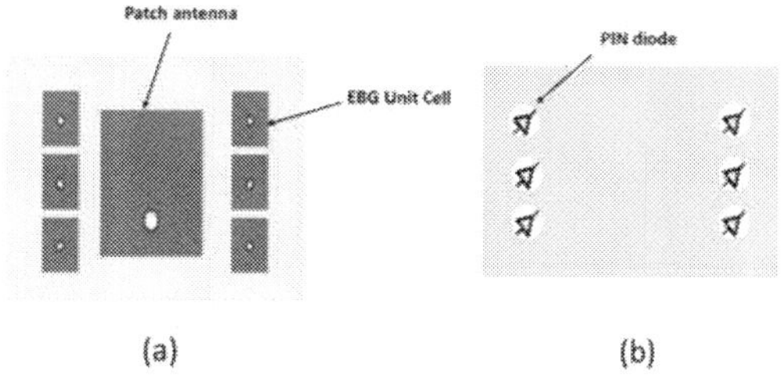

Figure 18. Reconfigurable pattern antenna using HIS structure: (a) top view and (b) bottom view.

Six PIN diodes are used as a switcher to control the surface-wave propagation according to the equation [3]

$$\frac{\beta}{k_0} = \sin(\theta) \tag{6}$$

where β is phase coefficient of complex constant propagation k, k_0 is the free space wavenumber, and θ is the steered angle of the radiation pattern.

As seen previously in Section 2, the HIS ground plane is divided in two parts: one part acts as a reflector for TE surface wave by activating the corresponding PIN diodes. While other PIN

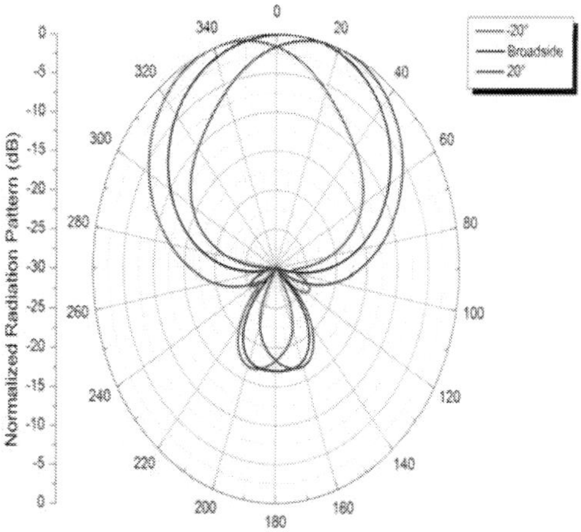

Figure 19. Small-cell normalized Azimuthal radiation patterns.

http://dx.doi.org/10.5772/67522

diodes are turned OFF, allowing the propagation of TM surface waves which radiate with patches printed on top of the substrate.

According to **Figure 19**, three steered beams inside a sector at −20°, 0°, and 20° are achieved, respectively.

By replacing the PRS of the hexagonal FPC LWA by the proposed active HIS, the whole system divides the Azimuth plane in 18 sectors, according to **Figure 20**, in simple ways using few PIN diodes [3].

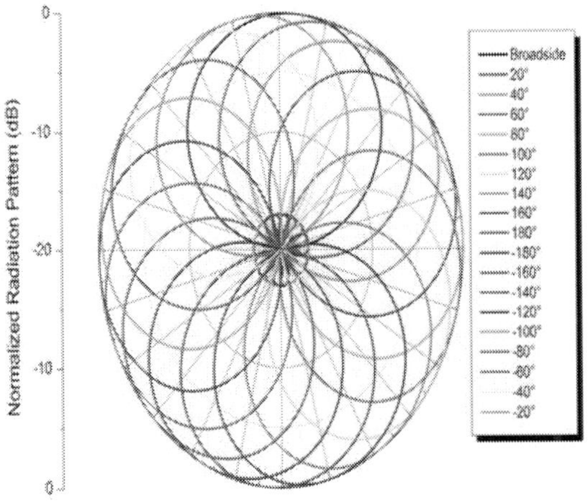

Figure 20. Eighteen beams covering azimuth plane.

4. Conclusion

SBA systems based on periodic metal structures have been presented in this chapter showing their ability in 360° of azimuthal coverage for base station applications. First, RPA has been studied which is composed of a linear slot-antenna array surrounded by an active FSS. In this design, PIN diodes are used as a switcher to control FSS response, giving an agility to omnidirectional radiation pattern and create SBA which divides the azimuth plane in six sectors. However, the major drawback of this system consists of a large number of active elements used which introduces losses and complicates the system's architecture.

For that, a hexagonal FPC LWA array has been proposed as solution to overcome problems related to PIN diodes. A high gain and good radiation efficiency are obtained from patch antenna, located inside a planar FPC, to avoid the use of antenna arrays with their complex feeding networks. An external SP6T RF switcher is required to be connected to FPC LWA arrays to switch the beam from a sector to another.

To increase more sectorization level, a reconfigurable HIS ground plane is proposed as a beamforming technique to divide the sector in three small cells. As a result, 18 beams are achieved covering all azimuth planes. The proposed SBA structure, based on FPC LWA, is the most suitable for future wireless base station applications due to its simplicity.

Author details

Mohamed Aymen El Cafsi[1]*, Mourad Nedil[2], Lotfi Osman[1] and Ali Gharsallah[1]

*Address all correspondence to: aymenelcafsi@gmail.com

1 Faculty of Sciences of Tunis, University of Tunis El Manar, UR13ES37 Unit of Research in High Frequency Electronics Circuits and Systems, Tunis, Tunisia

2 University of Quebec in Abitibi-Témiscamingue, Engineering School, Québec, Canada

References

[1] Habib M.A., Jazi M.N., Djaiz A., Nedil M., Denidni T.A. Switched-beam antenna based on EBG periodic. In: IEEE, editor. IEEE MTT-S International Microwave Symposium Digest; 7–12 June 2009; Boston. IEEE; 2009. DOI: 10.1109/MWSYM.2009.5165821

[2] Nedil M., Denidni T.A.,Talbi L. Novel butler matrix using CPW multilayer technology. IEEE Transactions on Microwave Theory and Techniques. 2006;**54**(1):499–507. DOI: 10.1109/TMTT.2005.860490

[3] El Cafsi M.A., Nedil M., Osman L., Gharsallah A. High sectorization switched beams for WLAN applications. In: IEEE International Symposium on Antennas and Propagation (APSURSI); 26 June–1 July; Puerto Rico. IEEE; 2016. pp. 2161–2162. DOI: 10.1109/APS.2016.7696787

[4] El Cafsi M.A., Nedil M., Osman L., Gharsallah A. An efficient hexagonal switched beam antenna structure based on Fabry-Pérot cavity leaky-wave antenna. International Journal of Electronics. 2015;**102**(11):1789–1803. DOI: http://dx.doi.org/10.1080/00207217.2014.996782

[5] Mosca S., Bilotti F., Toscano A., Vegni L. A novel design method for Blass matrix beam-forming networks. IEEE Transactions on Antennas and Propagation. 2002;**50**(2):225–232. DOI: 10.1109/8.997999

[6] Rotman W., Turner R.F. Wide-angle microwave lens for line source applications. IEEE Transactions on Antennas and Propagation. 1963;**11**(6):623–632. DOI: 10.1109/TAP.1963.1138114

[7] Christie S., Cahill R., Buchanan B., Fusco V.F., Mutchell N., Munro Y.V., Maxwell-Cox G. Rotman lens-based retro directive array. IEEE Transactions on Antennas and Propagation. 2012;**60**(3):1343–1351. DOI: 10.1109/TAP.2011.2180347

[8] Huang Y. Smart antenna systems for WiMAX radio technology. In: WIMAX New Developments. Rijeka: InTech; 2009. pp. 201–212. DOI: 10.5772/8269

[9] El Cafsi M.A., Nedil M., Osman L., Gharsallah A. A reconfigurable switched beam antenna using active cylindrical frequency selective surface. In: 26th IEEE Canadian Conference Of Electrical And Computer Engineering (CCECE); May 5–8; Regina, Canada. IEEE; 2013. pp. 1–4. DOI: 10.13140/RG.2.1.5094.3448

[10] Edalati A., Denidni T.A. High-gain reconfigurable sectoral antenna using an active cylindrical FSS structure. IEEE Transactions on Antennas and Propagation. 2011;**59**(7):2464–2472. DOI: 10.1109/TAP.2011.2152327

[11] Bouslama M., Traii M., Denidni T.A., Gharsallah A. Beam-switching antenna with a new reconfigurable frequency selective surface. IEEE Antennas and Wireless Propagation Letters. 2015;**15**:1159–1162. DOI: 10.1109/LAWP.2015.2497357

[12] COMMSCOPE. Six sector Solution [Internet]. 2013.Available from: http://www.commscope.com/Solutions/Six-Sector-Turnkey-Solution/

[13] Burghignoli P., Lovat G., Capolino F., Jackson D.R., Wilton D.R. High polarized, directive radiation from a Fabry-Pérot cavity leaky-wave antenna based on a metal strip grating. IEEE Transactions on Antennas and Propagation. 2010;**58**(12):3873–3883. DOI: 10.1109/TAP.2010.2078441

[14] Jackson D.R., Burghignoli P., Lovat G., Capolino F., Ji C., Wilton D.R., Oliner A.A. The fundamental physics of directive beaming at microwave and optical frequencies and the role of leaky waves. Proceeding of the IEEE. 2011;**99**(10):1780–1805. DOI: 10.1109/JPROC.2010.2103530

[15] Zhao T., Jackson D.R., Williams J.T., Oliner A.A. General formulas for 2-D leaky wave antennas. IEEE Transactions on Antennas and Propagation. 2005;**53**(11):3525–3533. DOI: 10.1109/TAP.2005.856315

[16] Lovat G., Burghignoli P., Capolino F., Jackson D.R., Wilton D.R. High gain omnidirectional radiations patterns from a metal strip grating leaky wave. In: IEEE International Symposium Antennas and Propagation Society; 9–15 June; Hawaii. IEEE; 2007. pp. 5797–5800. DOI :10.1109/APS.2007.4396869.

[17] El Cafsi M.A., Nedil M., Osman L., Gharsallah A. 1D switched beam Fabry-Pérot leaky wave antenna. In: IEEE Antennas and Propagation Society International Symposium (APSURSI); 2014; Memphis. IEEE; 2014. pp. 1798–1799. DOI: 10.1109/APS.2014.6905225

[18] Garcia-Vigueras M., DeLara-Guarch P., Gomez-Tomero J.L., Guzman-Quiros R., Goussetis G. Efficiently illuminated broadside-directed 1D and 2D tapered Fabry-Pérot leaky-wave antenna. In: 6th European Conference on Antenna and Propagation (EUCAP);2012; 26–30 March; Prague. IEEE; 2012. pp. 247–251. DOI: 10.1109/EuCAP.2012.6206359

[19] Gardelli R., Albani M., Capolino F. Array thinning by using antennas in a Fabry-Pérot cavity for gain enhancement. IEEE Transactions on Antennas and Propagation. 2006;**54**(7):1979–1990. DOI: 10.1109/TAP.2006.877172

[20] Komanduri V.R., Jackson D.R., Long S.A. Radiation characteristics of finite-length 1D-uniform leaky wave antennas radiating at broadside. In: IEEE International Symposium Antennas and Propagation Society; 11–17 July; Toronto. IEEE; 2010. DOI: 10.1109/APS.2010.5560977

[21] Jaewoo L., Chan Han J., Sungweon K., Choi C.A. A low-loss single-pole six throw switch based on compact RF MEMS switches. IEEE Transactions on Microwave Theory and Techniques. 2005;**53**(11):3335–3344. DOI: 10.1109/TMTT.2005.855746

[22] COMMSCOPE. Twin beam technology adds immediate capacity without additional antennas [Internet]. May 2013. Available from: http://www.commscope.com/Docs/Six-Sector_Twin_Beam_WP-106683.pdf

[23] Ansoft HFSS. High Frequency Electromagnetic Field Simulation [Internet]. Available from: http://www.ansys.com/Products/Electronics/ANSYS-HFSS

[24] CST Microwave Studio. CST MICROWAVE STUDIO [Internet]. Available from: https://www.cst.com/products/cstmws

Chapter 3

Compressive Reflector Antenna Phased Array

Ali Molaei, Juan Heredia Juesas and

Jose Angel Martinez Lorenzo

Additional information is available at the end of the chapter

http://dx.doi.org/10.5772/67663

Abstract

Conventional phased array imaging systems seek to reconstruct a target in the imaging domain by employing many transmitting and receiving antenna elements. These systems are suboptimal, due to the often large mutual information existing between two successive measurements. This chapter describes a new phased array system, which is based on the use of a novel compressive reflector antenna (CRA), that is capable of providing high sensing capacity in different imaging applications. The CRA generates spatial codes in the imaging domain, which are dynamically changed through the excitation of multiple-input-multiple-output (MIMO) feeding arrays. In order to increase the sensing capacity of the CRA even further, frequency dispersive metamaterials can be designed to coat the surface of the CRA, which ultimately produces spectral codes in near- and far- fields of the reflector. This chapter describes different concepts of operation, in which a CRA can be used to perform active and passive sensing and imaging.

Keywords: compressive antenna, spatial and spectral coding, reflector antenna, metamaterial absorbers, sensing capacity

1. Introduction

Reducing the cost of electromagnetic (EM) sensing and imaging systems is a necessity before they can be far and widely established as a part of an extensive network of radars. Recently, a new beamforming technique based on a compressive reflector antenna (CRA) was proposed [1–6] to improve the sensing capacity of an active sensing system. This improvement has enhanced the information transfer efficiency from the sensing system to the imaging domain and vice versa. Thus, complexity and cost of the hardware architecture can be drastically reduced. The beamforming that the CRA creates is based on multi-dimensional coding: (a) spatial coding by introducing dielectric or metallic scatterers on the surface of the reflector, (b) spectral coding by coating the reflector with metamaterials, and (c) temporal coding by the use of temporal multiplexing of transmitting and receiving horn arrays.

This unique feature of CRAs has triggered its use in a wide variety of applications, which include the following: (a) active imaging of metallic targets at mm-wave frequencies [1–3], (b) passive imaging of the physical temperature of the Earth at mm-wave frequencies [4, 5], and (c) active imaging of red blood cells at optical frequencies [6].

The proposed CRA beamforming technique, which may be used for imaging applications, is based on norm-1 regularized iterative Compressive Sensing (CS) algorithms. In this chapter, we also present the mathematical formulation that describes the properties of the spatial and temporal codes produced by the CRA that will be used to perform quasi real-time imaging.

The content outlined in this chapter leverages advances from multi-scale wave propagation, sparse data signal processing, information coding, and distributed computing. The result will enhance the efficiency and reliability of the current beamforming systems by using novel compressive sensors made of traditional metallic and dielectric structures, as well as novel metamaterials and meta-surfaces.

2. Compressive reflector antenna

The concept of operation of the CRA for sensing and imaging applications relies on two basic principles: (a) multi-dimensional coding, generated by the design of a customized reflector and (b) compressed sensing, performed on the under-sampled measured data.

The CRA is fabricated as described in Ref. [1]. **Figure 1** shows the cross-section of a traditional reflector antenna (TRA) (x > 0) and of a CRA (x < 0). The latter is built by introducing discrete scatterers, Ω_i, on the surface of the reflector, which are characterized by their size $\{D_i^x, D_i^y, D_i^z\}$ in $\{\hat{x}, \hat{y}, \hat{z}\}$, and by the electromagnetic constitutive properties $\{\sigma_i, \mu_i, \varepsilon_i\}$ (conductivity, permeability, and permittivity). These scatterers generate a pseudo-random time-space coded pattern in the near and far fields of the antenna after reflecting the incident field produced by the feeding elements. The temporal variation of these coded patterns allows the use of CS techniques to generate a 3D image of an object under test in the region of interest (ROI). The main difference between a CRA and a TRA is the use of the applique scatterers located on the surface of the former, as shown in **Table 1**.

2.1. Sensing matrix

There are many techniques that may be used to dynamically change these coded patterns, including but not limited to the following: (a) electronic beam steering by using a focal plane array or a reconfigurable sub-reflector, (b) electronic change of the constitutive parameters of the scatterers, and (c) mechanical rotation of the reflector along the $\hat{z}$-axis of the parabola.

http://dx.doi.org/10.5772/67663

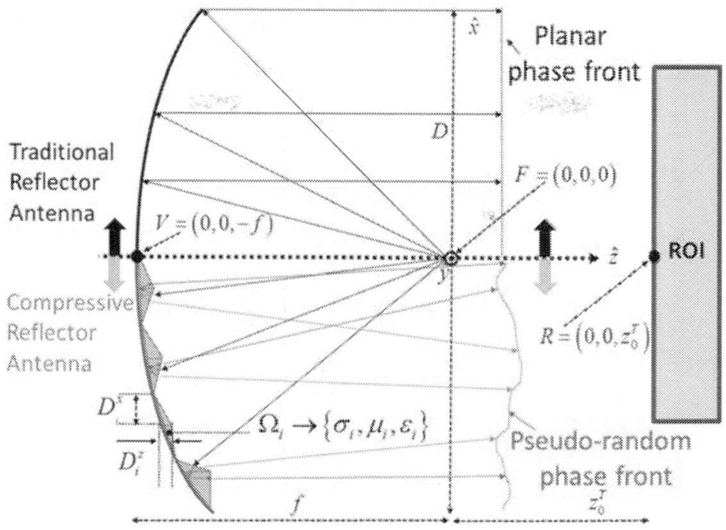

Figure 1. 2D cross-sections of a traditional reflector antenna ($x > 0$) and a compressive reflector antenna (CRA) ($x < 0$), where all the design parameters are described.

Parameter	TRA		CRA
D: aperture size		Same	
f: focal length		Same	
Ω_i: scatterer	No		Yes

Table 1. Comparison of the parameters of reflector antennas.

Let us focus on option 1, where an array of N_{Tx} transmitter and N_{Rx} receiver horns are arranged in a cross-shaped configuration around the focal point of the reflector, as shown in **Figure 2**. Each receiver collects the signal from each transmitter for N_f, different frequencies, for a total number of $N_m = N_{Tx} \cdot N_{Rx} \cdot N_f$ measurements. The image reconstruction is performed in N_p pixels, on an region of interest (ROI) located z_0^T meters away from the focal point of the CRA. Under this configuration, the sensing matrix $\mathbf{H} \in \mathbb{C}^{N_m \times N_p}$, computed as described in Ref. [7], establishes a linear relationship between the unknown complex reflectivity vector in each pixel, $\mathbf{u} \in \mathbb{C}^{N_p}$, and the measured complex field data $\mathbf{g} \in \mathbb{C}^{N_m}$. This relationship can be expressed in a matrix form, by applying the physical optics (PO) approximation to the total field and discretizing the integral operator as done in Ref. [8], as follows:

$$\mathbf{g} = \mathbf{H}\mathbf{u} + \mathbf{w}, \tag{1}$$

where $\mathbf{w} \in \mathbb{C}^{N_m}$ represents the noise collected by each receiving antenna, when the target is illuminated with a given transmitting antenna and for a given frequency.

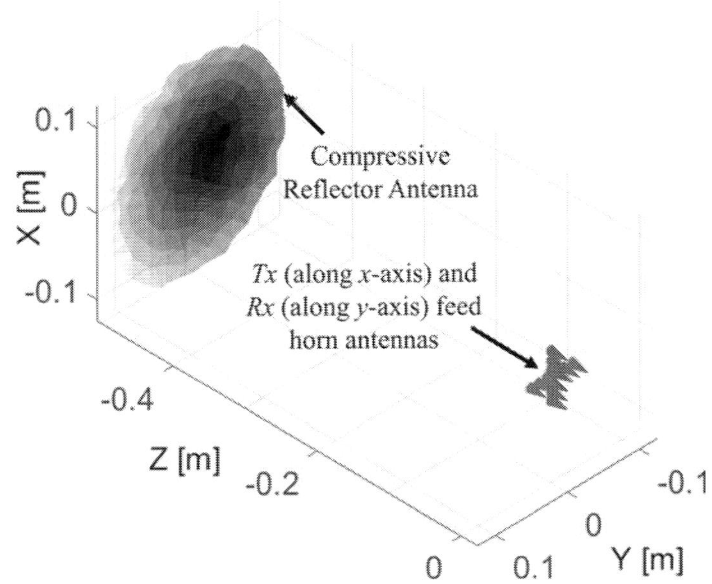

Figure 2. Geometry of the compressive reflector antenna with the feed horns at the focal plane.

Regardless of the configuration of the system, the sensing matrix will always have the dimensions $N_m \times N_p$, where N_m is the number of independent codes that are used—it depends on the configuration and the type of imaging system—and N_p is the total number of pixels in the imaging domain.

In order to impose sparsity on the solution of Eq. (1), a compressive sensing (CS) approach is used. CS theory was first introduced by Candes et al. [9], and it establishes that sparse signals can be recovered by the use of a reduced number of measurements when compared to those required by the Nyquist sampling criterion. In order to be able to apply such principles, the sensing matrix **H** must satisfy the Restricted Isometry Property (RIP) condition [10], which is related to the independence of its columns. Likewise, the number of nonzero values N_{nz} of the reconstructed image vector **u** must be much smaller than the total number of elements N_p (that is $N_{rz} \ll N_p$). Under the assumption that the two aforementioned conditions are satisfied, the reconstruction of the unknown vector **u** may be performed by solving the following optimization problem that only uses a reduced number of measurements **g**:

$$\min\|\mathbf{u}\|_1 \quad \text{s.t.} \quad \|\mathbf{Hu} - \mathbf{g}\|_2 < \delta_H \tag{2}$$

where δ_H is an upper bound for the residual error $\|\mathbf{Hu} - \mathbf{g}\|_2$. Many algorithms for solving Eq. (2) have been developed [11, 12]. Here, a MATLAB toolbox NESTA [23] is used to solve that equation.

http://dx.doi.org/10.5772/67663

2.2. Sensing capacity of a compressive reflector antenna

The linearized sensing matrix **H** in Eq. (1) defines the properties of the imaging system, which can be interpreted as a multiple-input-multiple-output (MIMO) communications system [13]. The capability of the CRA to transmit information from the image domain **r** to the measured field domain **g**, in the presence of noise, can be studied by quantifying the capacity associated with the sensing matrix **H**. This sensing capacity of the imaging system can be derived from the singular value decomposition of **H**, as follows:

$$\mathbf{H} = \mathbf{U}\Sigma\mathbf{V}^{\star} \tag{3}$$

where $\mathbf{V} = (\mathbf{v}_1, \ldots, \mathbf{v}_{N_p})$ and $\mathbf{U} = (\mathbf{u}_1, \ldots, \mathbf{u}_{N_m})$ are, respectively, $N_p \times N_p$ and $N_m \times N_m$ matrices containing a set of orthonormal input- and output- based directions for **H**; the matrix $\Sigma = diag(\lambda_1, \ldots, \lambda_{N_{min}})$, where $N_{\min} = min(N_m, N_p)$, is an $N_p \times N_m$ matrix containing the real nonzero singular values of **H** in the diagonal and zeros elsewhere. When the *l–th* input base direction v_l is used in the image domain and propagated through the channel **H**, a $\lambda_l \mathbf{u}_l$ response is generated in the output-measured field domain. Therefore, $\{\lambda_l, \mathbf{v}_l, \mathbf{u}_l\}$ can be seen as the parameters of the *l–th* orthogonal channel of the matrix **H**. The $N_{\min}$ orthogonal parallel channels provide the following capacity, measured in bits [13]:

$$C = \sum_{l=1}^{N_{min}} \log_2\left(1 + \frac{P_l \lambda_l^2}{N_0}\right) \tag{4}$$

where P_l/N_0 is the signal to noise ratio (SNR) in the *l–th* orthogonal channel. The parameters $\{\sigma_i, \mu_i, \epsilon_i, D_i^x, D_i^y, D_i^z\}$ are used to modify and control the singular values of the matrix **H** and, therefore, to tailor the sensing capacity of the imaging system. One important feature that is often desired for a sensing and an imaging system is its ability to maximize the information transfer efficiency, that is, its sensing capacity, between the pixels in the region of interest and the sensors; this happens when the mutual information of successive measurements is reduced as much as possible. The pseudo-random spatial codes created by the CRA make successive measurements more independent, which ultimately results in measurements having reduced mutual information and providing enhanced imaging capabilities to the system.

2.3. Metamaterial absorber-based compressive reflector antenna

A metamaterial absorber (MMA) [4, 5, 13–15] , which was originally introduced by Landy et al. [15], poses a unique behavior that can be exploited for sensing and imaging applications. Specifically, by using an array of MMAs, in which each element of the array presents a near-unity absorption at a specified frequency, one can produce codes that are changed with the instantaneous frequency of the radar chirp, as presented in Ref. [16]. As a result, the number of transmitters and receivers required to achieve suitable imaging performance is drastically reduced. Coating the recently developed CRA with MMAs has the potential to further improve the antenna's imaging capabilities, in terms of sensing capacity (**Figure 3**) [1]. However, the utilization of the MMAs in doubly curved pseudo-randomly distorted compressive reflectors for imaging applications requires an accurate characterization of the bulk behavior of

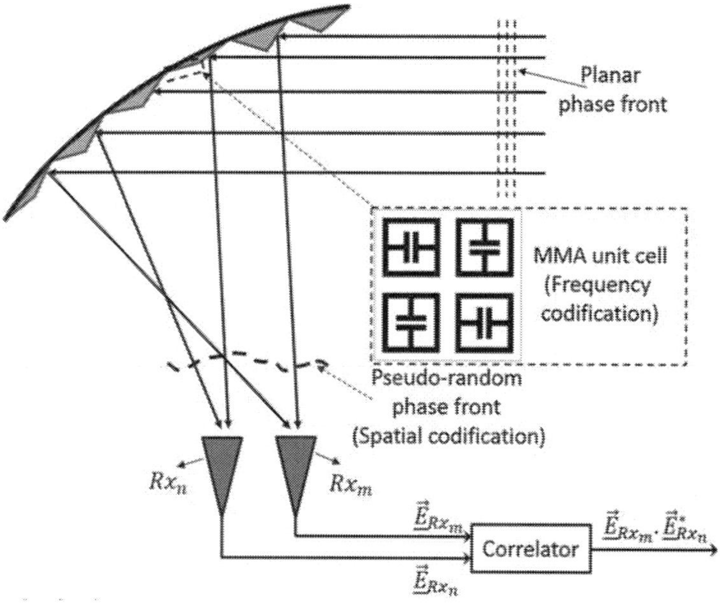

Figure 3. A 2D cross-section of an offset metamaterial-based CRA.

the metamaterial for dimensional scales involving several wavelengths and for oblique incidence on the MMAs. The MMA array can be characterized by solving a three-layer magneto-dielectric medium problem, where an incident field is obliquely impinging a magneto-dielectric medium of thickness *d*, which is backed by a metallic layer. The magneto-dielectric and metallic layers may be characterized by a Drude-Lorentz model [14]:

$$\tilde{\varepsilon}_r(\omega) = \varepsilon_{\text{inf}} + \frac{\omega_{p,e}^2}{\omega_{0,e}^2 - \omega^2 - i\gamma_e\omega}; \qquad \tilde{\mu}_r(\omega) = \mu_{\text{inf}} + \frac{\omega_{p,m}^2}{\omega_{0,m}^2 - \omega^2 - i\gamma_m\omega} \tag{5}$$

in which, ε_{inf} and μ_{inf} are the static permittivity and permeability at infinite frequency, $\omega_{p,e}$ and $\omega_{p,m}$ are the volume plasma frequencies at which the density of the electric and magnetic charges oscillate, $\omega_{0,e}$ and $\omega_{0,m}$ are the resonant frequencies, and γ_e and γ_m are the damping constants, which represent the electric and magnetic charge collision rate.

The reflection coefficient of this stratified three-layer magneto-dielectric medium can be analytically described as follows [17]:

$$\Gamma = \Gamma_{12} + \frac{T_{12}\Gamma_{23}T_{21}e^{-j2\Phi^{trans}}}{1 - \Gamma_{23}\Gamma_{21}e^{-2j\Phi^{trans}}} \tag{6}$$

where Γ is the total reflection coefficient of the structure, Γ_{ij} and T_{ij} are the reflection and transmission coefficients, associated with the interface between medium i and medium j, respectively, Φ^{trans} is the phase delay and the amplitude attenuation associated with the wave

http://dx.doi.org/10.5772/67663

traveling from the first interface into the second one, or vice versa, and $e^{-j\Phi^{trans}} = e^{-jk_2^t \cdot r} = e^{-j\beta_2^t \cdot r} \cdot e^{-\alpha_2^t \cdot r}$, with **r** being the distance vector and k_2^t being the complex wave vector. Although this model does not take anisotropy into consideration, it can be included in the model using the formulation described in Ref. [18].

2.4. Beamforming using compressive reflector antenna

When a TRA is illuminated from the focus (located at $z = 0$, without loss of generality, as **Figure 1** shows), a plane wave field is obtained in the focal aperture. The mathematical expression of this wave may be represented in terms of both spatial and time coordinates as follows:

$$g(x, y, z, t) = g_0(x, y, z)e^{j\omega_0 t}, \tag{7}$$

where $\omega_0 = 2\pi f_0$, with f_0 being the frequency of the wave. The spatial distribution of the field $g_0(x, y, z)$ can be split as $g_0(x, y, z) = A(z)I_\Omega(x, y)$, with $A(z)$ as the amplitude, which depends on the z position, and $I_\Omega(x, y)$ as an indicator function defined as follows:

$$I_\Omega(x, y) = \begin{cases} 1 & if \quad (x, y) \in \Omega \\ 0 & \text{otherwise,} \end{cases} \tag{8}$$

Ω being the domain of definition of the antenna, that is, the 2D projection of the reflector in the plane XY.

The radiation pattern in the far-field is related to the field in the aperture through a 2D Fourier transform as follows [19]:

$$\begin{aligned} G(\theta, \phi, z, t) = \mathbf{FT}[g(x, y, z, t)] &= \int_\Omega g_0(x, y, z)e^{j\omega_0 t}e^{jk_u x}e^{jk_v y}dxdy \\ &= A(z)e^{j\omega_0 t}\int_\Omega I_\Omega(x, y)e^{jk_u x}e^{jk_v y}dxdy = A(z)e^{j\omega_0 t}\mathbf{FT}[I_\Omega(x, y)], \end{aligned} \tag{9}$$

for $u = \sin(\theta)\cos(\phi)$ and $v = \sin(\theta)\sin(\phi)$.

In the case where applique scatterers are added to the surface of the reflector, so creating the compressive reflector antenna, the field in the focal aperture is not going to be uniform any more. It can be defined as follows:

$$\tilde{g}(x, y, z, f, t) = \tilde{g}_0(x, y, z, f)e^{j\omega_0 t}. \tag{10}$$

The new expression of the field depends as well on the frequency, and its spatial distribution can be expressed as follows:

$$\tilde{g}_0(x, y, z, f) = g_0(x, y, z)c(x, y, f). \tag{11}$$

The function $c(x, y, f)$ describes the changes on the focal aperture field with respect to that of a traditional reflector, due to the random rugosity in space and the random variation in frequency associated with the dispersive metamaterials. These functions are interpreted as codes.

The radiation pattern would now be determined, as well, in terms of the Fourier transform as follows [20]:

$$\begin{aligned}\tilde{G}(\theta,\phi,z,t,f) &= \mathbf{FT}[\tilde{g}(x,y,z,t,f)] = \int_{\Omega} g_0(x,y,z)c(x,y,f)e^{j\omega_0 t}e^{jk_u x}e^{jk_v y}dxdy \\ &= A(z)e^{j\omega_0 t}\int_{\Omega} I_{\Omega}(x,y)c(x,y,f)e^{jk_u x}e^{jk_v y}dxdy = \frac{1}{2\pi}A(z)e^{j\omega_0 t}\mathbf{FT}[I_{\Omega}(x,y)]{*}{*}\mathbf{FT}[c(x,y,f)].\end{aligned} \tag{12}$$

In this way, the beamforming is performed by the 2D spatial convolution of the original pattern with the Fourier transform of the code. **Figure 4** shows the difference between the radiation pattern of a traditional reflector and that of a compressive reflector in one dimension.

In the general case, when an array of transmitters and receivers is arranged around the focal point of the reflector (**Figure 2**), the two-way radiation pattern for a traditional reflector is given by the product of the transmitting and receiving radiation patterns, $G_{Tx}(\theta,\phi,z,t)$ and $G_{Rx}(\theta,\phi,z,t)$, respectively [19]. These are defined by the Fourier transform of both the transmitter and receiver apertures, $g_{Tx}(x,y,z,t)$ and $g_{Rx}(x,y,z,t)$, respectively:

$$\begin{aligned}G_{TR}(\theta,\phi,z,t) &= G_{Tx}(\theta,\phi,z,t)G_{Rx}(\theta,\phi,z,t) \\ &= \mathbf{FT}[g_{Tx}(x,y,z,t)]\mathbf{FT}[g_{Rx}(x,y,z,t)] = \mathbf{FT}[g_{Tx}(x,y,z,t){*}{*}g_{Rx}(x,y,z,t)].\end{aligned} \tag{13}$$

In the case of the compressive reflector antenna, the two-way radiation pattern will be determined by the product of the transmitting and receiving radiation patterns, $\tilde{G}_{Tx}(\theta,\phi,z,t,f)$ and $\tilde{G}_{Rx}(\theta,\phi,z,t,f)$, respectively, considering, as well, the codes in space and frequency:

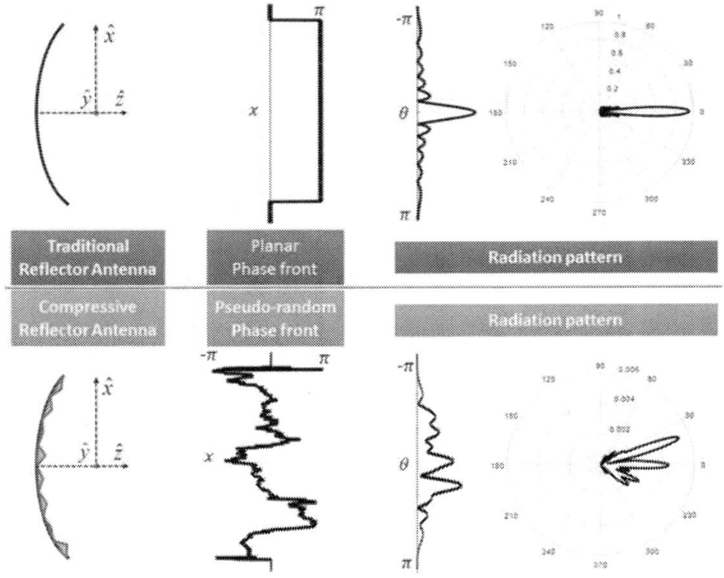

Figure 4. Top: A 2D cross-section of a traditional reflector antenna, planar phase front distribution, and a radiation pattern with a main lobe. Bottom: A 2D cross-section of a compressive reflector antenna, pseudo-random phase front distribution, and a radiation pattern with multiple lobes.

http://dx.doi.org/10.5772/67663

$$
\begin{aligned}
\tilde{G}_{TR}(\theta,\phi,z,t,f) &= \tilde{G}_{Tx}(\theta,\phi,z,t,f)\tilde{G}_{Rx}(\theta,\phi,z,t,f) \\
&= \mathbf{FT}[\tilde{g}_{Tx}(x,y,z,t,f)]\mathbf{FT}[\tilde{g}_{Rx}(x,y,z,t,f)] = \mathbf{FT}[\tilde{g}_{Tx}(x,y,z,t,f)**\tilde{g}_{Rx}(x,y,z,t,f)] \\
&= \mathbf{FT}[g_{Tx}(x,y,z,t)c_{Tx}(x,y,f)**g_{Rx}(x,y,z,t)c_{Rx}(x,y,f)] \\
&= \frac{1}{4\pi^2}\Big(\mathbf{FT}[g_{Tx}(x,y,z,t)]**\mathbf{FT}[c_{Tx}(x,y,f)]\Big)\Big(\mathbf{FT}[g_{Rx}(x,y,z,t)]**\mathbf{FT}[c_{Rx}(x,y,f)]\Big) \\
&= \frac{1}{4\pi^2}\Big(G_{Tx}(\theta,\phi,z,t)**C_{Tx}(\theta,\phi,f)\Big)\Big(G_{Rx}(\theta,\phi,z,t)**C_{Rx}(\theta,\phi,f)\Big),
\end{aligned}
\tag{14}
$$

where $C_{Tx/Rx}(\theta,\phi,f) = \mathbf{FT}[c_{Tx/Rx}(x,y,f)]$. That is, the radiation pattern of transmitting and receiving arrays is convolved with the 2D spatial Fourier transform of the transmitting and receiving codes, respectively, and then multiplied. In this way, the beamforming is performed. Examples of radiation patterns for different CRA configurations are shown in **Figure 5**. The abovementioned formulations are combined with the method of moments in order to compute the scattered fields. The simulation results for a $15\lambda \times 15\lambda \times 2\lambda$ metallic, dielectric, and metamaterial bulk scatterers are presented in **Figure 5**. These simulations show the feasibility of using the CRA in order to generate wide pseudo-random sub-beam-like codes. These codes are produced not only in the near-field region but also in the far-field region of the CRA.

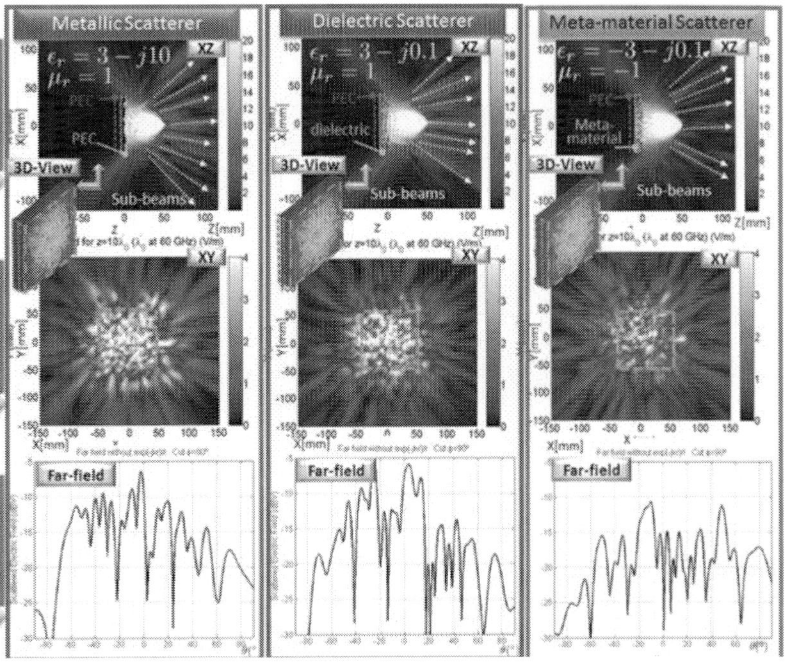

Figure 5. Multi-lobe radiation pattern representation for different scatterer configurations of a CRA.

The aforementioned electronic beamforming technique has been analyzed using a focal plane array on the proposed CRA. Point Spread Function (PSF) of the array is studied to measure the focusing efficiency of the system. The PSF can be evaluated by applying a phase compensation method. Specifically, the phase produced by each transmitting and receiving code of the CRA in the imaging region is adjusted in order to produce a zero phase at any desired focusing point. As a result, a constructive interference is obtained after adding all the codes that are in-phase at the focusing point, and a destructive interference is obtained elsewhere. The phased array consists of 12 equidistant transmitters on a vertical line on the focal plane, while the receiver elements are located similarly in a horizontal line. Eighteen frequencies are used in the range of 70–77 GHz to perform the beamforming, and the reflector has a diameter of 35 cm and a focal distance of 35 cm. **Figures 6** and **7** show the

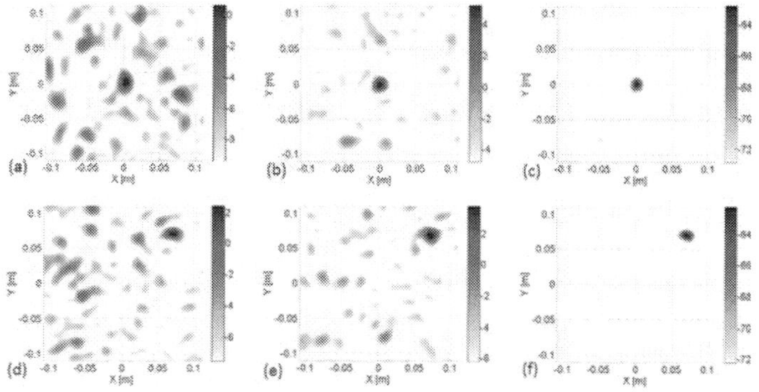

Figure 6. (a) PSF of the CRA with PEC scatterers. The focusing point is at [0, 0, 84] cm for (a), (b), and (c) and at [7, 7, 84] cm for (d), (e), and (f). (a) and (d) are the E-field patterns of the transmitters, (b) and (e) are the E-field patterns of the receivers, and (c) and (f) are the product of the E-field patterns of the transmitters and receivers.

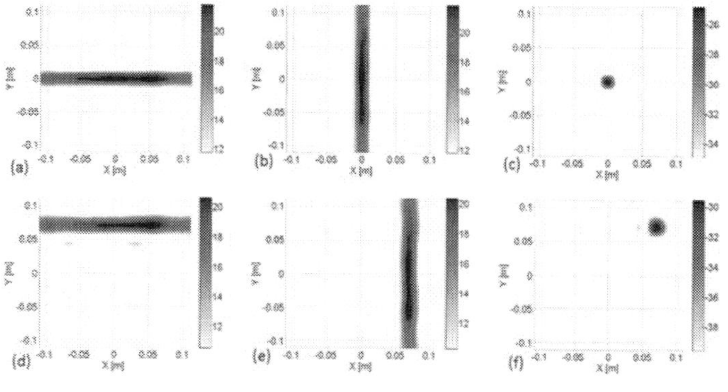

Figure 7. PSF of the CRA with MMAs. The focusing point is at [0, 0, 84] cm for (a), (b), and (c) and at [7, 7, 84] cm for (d), (e), and (f). (a) and (d) are the E-field patterns of the transmitters, (b) and (e) are the E-field patterns of the receivers, and (c) and (f) are the product of the E-field patterns of the transmitters and receivers.

simulated PSF of the CRA with perfect electric conductor (PEC) scatterers and MMAs, respectively. The triangle size of the PEC scatterers is 5λ, and MMAs are used to produce 16 different codes in the frequency domain. The imaging plane is located 84 cm far from the focal point.

3. Applications

In this section, the performance of the CRA for active and passive imaging applications is studied and compared to that of conventional systems. In all examples, the sensing capacity and image reconstruction of the systems are presented. It has been shown that the sensing capacity of the CRA is improved when compared to that of conventional imaging systems, and, as a result, a better image reconstruction can be achieved.

3.1. Active imaging

The performance of the CRA is evaluated in an active imaging application [1]. For this experiment, we use a mechanical rotation of the reflector along the $\hat{z}$-axis of the parabola, from 0 to θ_r degrees in N_θ steps.

Each scatterer Ω_i of the CRA is made of a PEC ($\sigma_i = \sigma_{PEC}$). The CRA is discretized into triangular patches, as described in Ref. [7], that are characterized by an average size of $< D^x >=< D^y >$ in $\hat{x}$ and $\hat{y}$ dimensions. The parameter λ_c is the wavelength at the center frequency. The scatterer size D_i^z of each triangle in $\hat{z}$ is modeled as a uniform random variable. The ROI is located at z_0^T away from the focal point of the CRA, and it encloses a volume determined by the following dimensions: Δx_0^T in $\hat{x}$, Δy_0^T in $\hat{y}$, and Δz_0^T in $\hat{z}$. The ROI is discretized into cubes of side length l. The total number of measurements (rows in **H**) is $N_m = N_\theta \cdot N_f$, and the total number of pixels for the imaging reconstruction (columns in **H**) is N_p. The parameters described in **Table 2** are used for the numerical simulation.

Three different configurations are analyzed in this example: (a) a TRA without scatterers on its surface, (b) a CRA with a feeding horn located in the focal point of the reflector (CRA-in-focus), and (c) a CRA with a feeding horn displaced $\Delta R^X = 10\lambda_c\hat{x} = 0.05\hat{x}\ m$ off the focal point of the reflector (CRA-off-focus). **Figure 8** shows the structure of the CRA, and **Figure 9** depicts the spatial codes generated by the CRA-off-focus antenna for five different rotation angles in a 2D plane of the ROI.

Figure 10(a) shows the singular values for the three configurations. The TRA presents only three singular values greater than -50 dB, and, as a result, its capacity is reduced when compared to that provided by any of the CRA configurations. Rotation of the CRA antenna makes the off-focus configuration illuminate different sections of the reflector, thus producing different spatial codes in the region of interest; the CRA-in-focus illuminates the same spatial region of the CRA when it is rotated around its axis. This methodology makes the CRA-off-focus have a singular value distribution with less dispersion than the CRA-in-focus, which

Parameters	Value	Parameters	Value
Center frequency (f_c)	$60GHz$	No. of measurements (N_m)	93
Wavelength at center frequency (λ_c)	$5mm$	No. of pixels in ROI (N_p)	25000
Bandwidth	$6GHz$	$< D^x > = < D^y >$	$1.5\lambda_c$
Reflector diameter (D)	$100\lambda_c = 0.5m$	D_i^z	$U(\mp 0.54\lambda_c)$
Focal length (f)	$100\lambda_c = 0.5m$	z_0^T	$194.87\lambda_c$
Maximum rotation (θ_r)	90^0	$\Delta x_0^T = \Delta y_0^T$	$36\lambda_c$
No. of rotations (N_θ)	31	Δz_0^T	$7.5\lambda_c$
No. of frequencies (N_f)	3	Length size of cubes l	$1.5\lambda_c$

Table 2. Parameters of the numerical design.

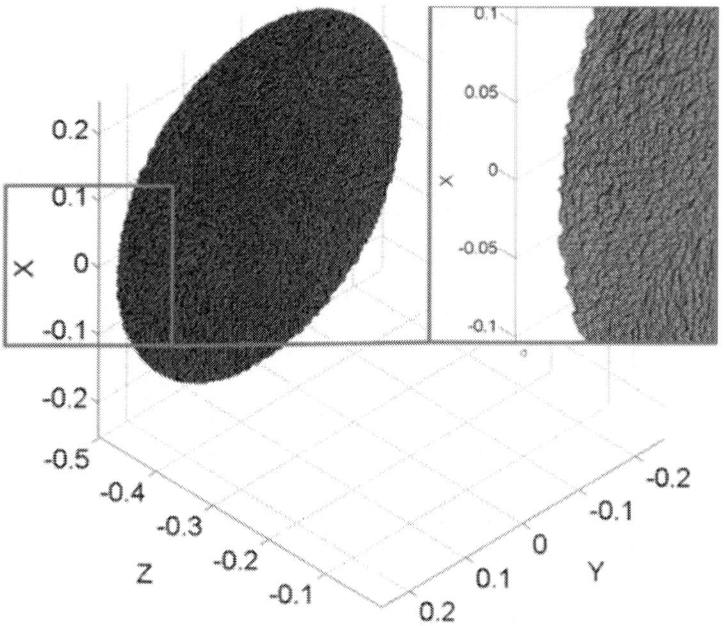

Figure 8. A 3D view of the CRA (left), and an augmented view of the pseudo-random scatterers (right).

ultimately provides the highest capacity of the three configurations in Eq. (4). **Figure 10(b)** shows the sensing capacity of the three configurations for different signal to noise ratios, and it can be seen that the CRA-off-focus clearly outperforms the other two configurations.

http://dx.doi.org/10.5772/67663

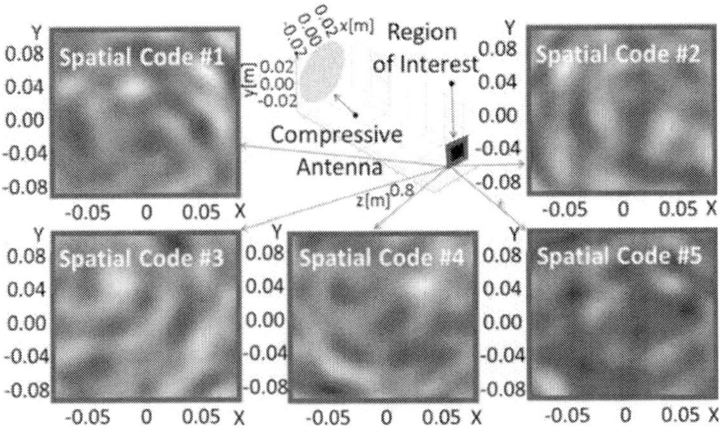

Figure 9. CRA and spatial codes generated in a 2D plane of the ROI for five rotation angles.

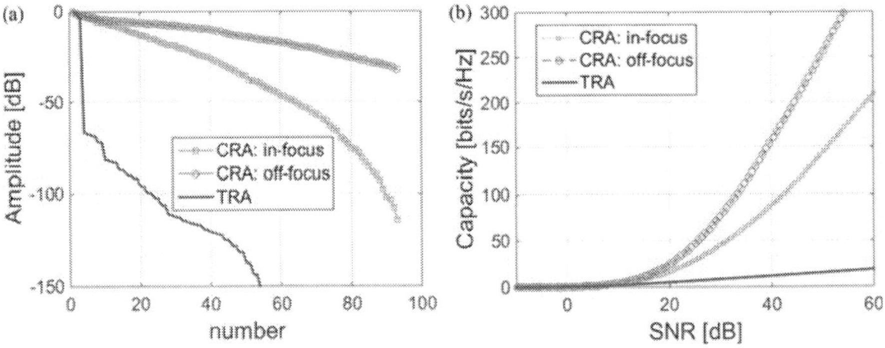

Figure 10. (a) Singular values in a logarithmic scale for the three configurations and (b) capacity as a function of the signal to noise ratio for the three configurations.

Figure 11 shows the imaging results. A uniform white noise producing a signal to noise ratio of 25 dB is used in the simulation. The target is represented by the transparent triangles with the black border. **Figure 11(a)** shows that, albeit CS is used, the sensing capacity of the in-focus CRA is not enough, and it fails to recover all the targets in the scene. However, the proposed off-focus CRA configuration is able to reconstruct objects with a sparsity level similar to that shown in **Figure 11(b)**. Notwithstanding, increasing the number of targets in the ROI may require additional measurements, which may be obtained from data collected at additional frequencies and/or rotation angles. The additional data results in an increment on the number of rows in the sensing matrix.

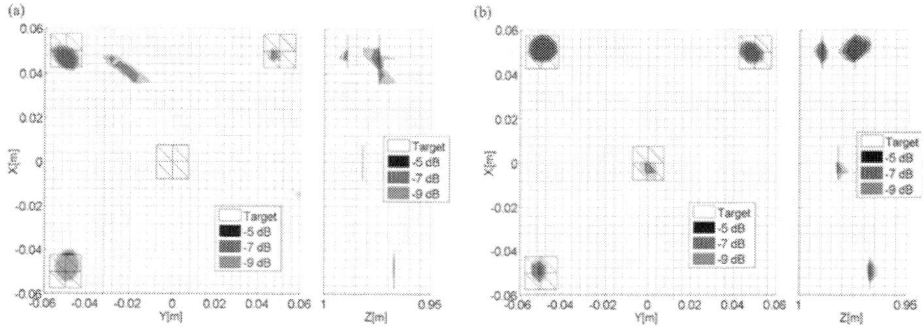

Figure 11. A front and a side view of the reconstructed normalized reflectivity function of the high sensing capacity using CS for (a) in-focus and (b) off-focus CRAs.

3.2. Array of CRAs for active imaging

In this example, the imaging system is composed of six CRAs positioned in a cross-shaped configuration, as shown in **Figure 12(a)**, each with an array of transmitters and receivers. The design parameters for each one of the reflectors are shown in **Table 3**. Both the vertical receiving array and the horizontal transmitting array of each CRA consist of 18 uniformly distributed conical horn antennas as shown in **Figure 12(b)**. The radar operates in the 70–77 GHz frequency band, and 10 frequencies are used to perform the imaging.

Each CRA is designed to effectively be able to image over a projected circular area of 40 cm diameter in the cross range region (see solid-line circles in **Figure 12(a)**) when the target is located 90 cm away from the focal plane. It is important to note that additional shaping

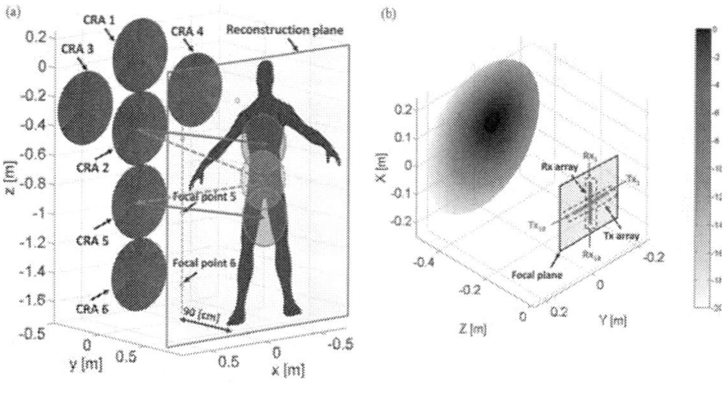

Figure 12. (a) A 3D view of the proposed millimeter-wave sensing system composed of six CRAs and (b) induced currents on a CRA excited by Tx1. The feeding transmitter (along y-axis) and receiver (along x-axis) arrays are located on the focal plane of the reflector.

Parameters	Value	Parameters	Value
Frequency band	70–77 GHz	No. of Tx	18
No. of frequencies (N_f)	10	No of Rx	18
Reflector diameter (D)	50 cm	$\langle D^x \rangle = \langle D^y \rangle$	2 cm
Focal length (f)	50 cm	D_i^z	Uniform (−10.5, +10.5) mm
Range (z_0^T)	90 cm		

Table 3. Design parameters for a single CRA.

techniques could have been used to image over a wider projected cross range region. The CRA has an aperture size of 50 cm, and, as a result, none of the two adjacent CRAs will be able to image the region located between their two circular projections (see the dashed-line circle in **Figure 12(a)**). This drawback can be easily solved by coupling the information coming from the adjacent reflectors in a multi-static fashion, as illustrated by the two dashed-line arrows in **Figure 12(a)**. Given the aforementioned location of the target, this work only considered the electromagnetic cross-coupling between CRA-l and CRA-k, where l and k take the following values: (l = 1, k = 2), (l = 1, k = 3), (l = 1, k = 4), (l= 2, k = 5), and (l = 5, k = 6).

The performance of the proposed active system is evaluated in a mm-wave imaging application, using a PO method [7]. The target used in the simulation is a tessellated model of a human body. In this work, the 3D human model was projected into a 2D plane, located 90 cm away from the focal plane, and its extension to 3D will be a future line of investigation. **Figure 13(a)** shows the improved singular value distribution of a single CRA when compared to that of a TRA, and **Figure 13(b)** shows how the sensing capacity of the CRA is enhanced for different SNRs.

Finally, **Figure 14** demonstrates that the proposed imaging system is capable of accurately reconstructing the target under investigation (note that the hands are out of the region of interest).

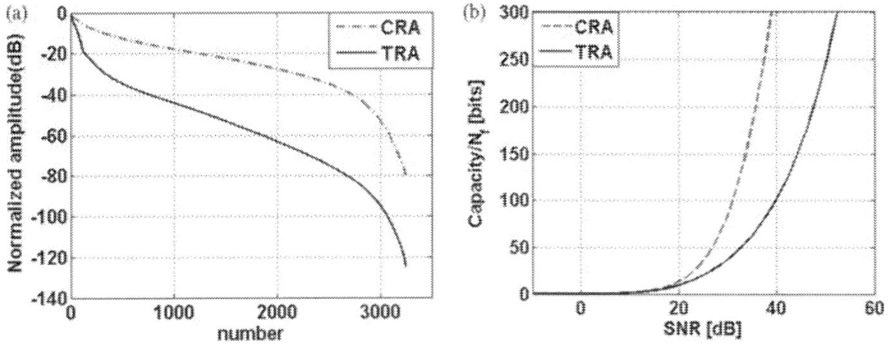

Figure 13. Comparison of (a) the normalized singular value distribution and (b) the sensing capacity of a single CRA and TRA.

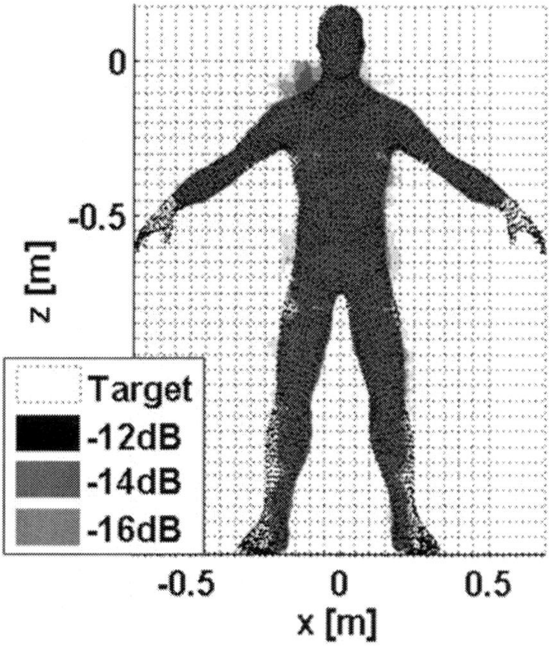

Figure 14. A reconstructed image using an iterative compressive sensing algorithm (NESTA).

3.3. Passive imaging

Using the mechanism introduced in Section 2.3, by describing the magneto-dielectric medium with the Drude-Lorentz model, three MMA types (MMA1, MMA2, and MMA3), resonating at three different frequencies (50 GHz, 52 GHz, and 54 GHz), are designed and randomly coated on the surface of the PEC scatterers, as shown in **Figure 15**. The polarization-independent electric-field-coupled absorber (ELCA) is designed using the commercially available software high-frequency structural simulator (HFSS)—a finite element-based full-wave solver [21]. Built-in master/slave boundary conditions in HFSS were utilized to simulate the planar MMA unit cell with periodic boundary conditions. Next, the Drude-Lorentz parameters of the three-layer magneto-dielectric medium are optimized to match the reflection coefficient of the model and the one obtained from HFSS for a given incident angle. The pattern search method embedded in the MATLAB optimization toolbox was used to solve the optimization problem. The optimized Drude-Lorentz parameters for MMA2 resonating at 52 GHz are as follows: $\varepsilon_{\text{inf}} = 2.9$, $\omega_{p,e} = 2\pi \times 9.01\, GHz$, $\omega_{0,e} = 2\pi \times 52\, GHz$, $\gamma_e = 2\pi \times 341\, MHz$, $\mu_{\text{inf}} = 3.1$, $\omega_{p,m} = 2\pi \times 7.55\, GHz$, $\omega_{0,m} = 2\pi \times 52\, GHz$, and $\gamma_m = 2\pi \times 291\, MHz$.

The performance of the designed metamaterial-based CRA interferometric system is evaluated in a microwave sounding imaging application, and it is then compared to that of a conventional interferometric system (GeoSTAR). The GeoSTAR system [22] is an interferometer, and its operation is based on performing complex cross-correlations between the measured fields by

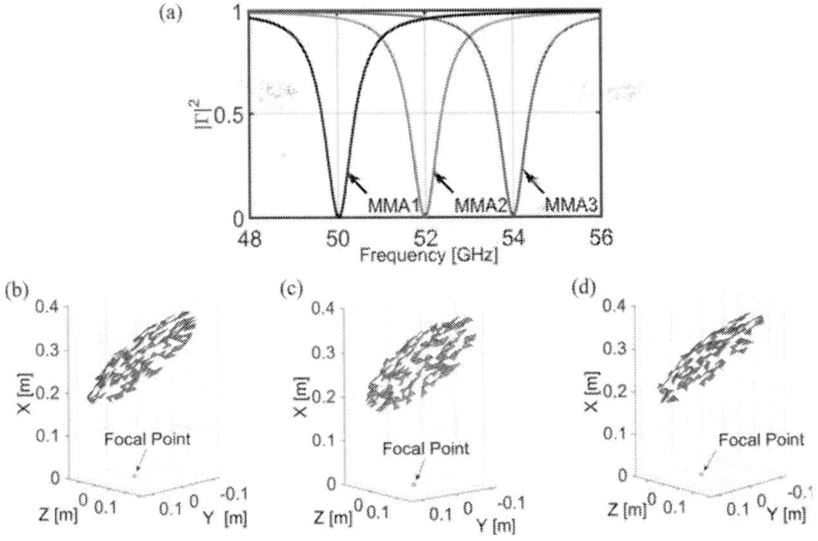

Figure 15. (a) Magnitude of the reflection coefficient and PEC scatterer facets associated with (b) MMA1, (c) MMA2, and (d) MMA3.

each pair of receivers in a Y-shaped array. These complex cross-correlated signals, which are characterized by the spatial coherence function of the electromagnetic field, are used to reconstruct the physical temperature of the Earth's atmosphere. For solving the inverse problem, a traditional pseudo-inverse method and a current state-of-the-art compressive sensing algorithm (NESTA) [23] are used.

The design parameters used for the numerical simulation are shown in **Table 4**. Nine receiving horns, which are placed in a Y-shaped configuration on the focal plane, are used to feed the metamaterial-based CRA. **Figure 16** shows a comparison of the geometry of the metamaterial-based CRA and GeoSTAR systems. The original image (**Figure 17(a)**) is an example of the physical temperature radiated from the surface of the Earth, and the system is assumed to measure EM fields from a geostationary satellite orbiting around the Earth. To ensure a fair

Parameters	CRA configuration	GeoSTAR configuration
Frequency band	50–54 GHz	50–54 GHz
Number of frequencies (N_f)	7	7
Longest Aperture size (D)	25 cm	25 cm
Diameter of feed elements	2.1 cm	2.1 cm
Number of feeds (N_r)	9	18
Focal length (f)	14 cm	–
Offset height (h_o)	28 cm	–

Table 4. Parameters for the numerical design.

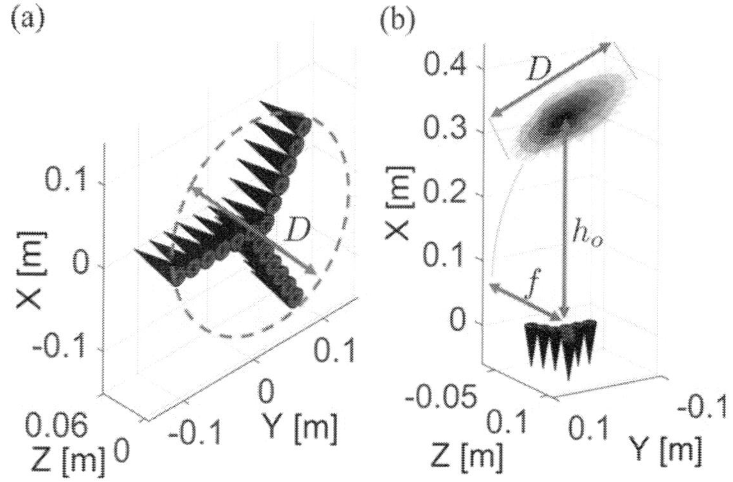

Figure 16. Geometry: (a) GeoSTAR configuration and (b) compressive reflector antenna.

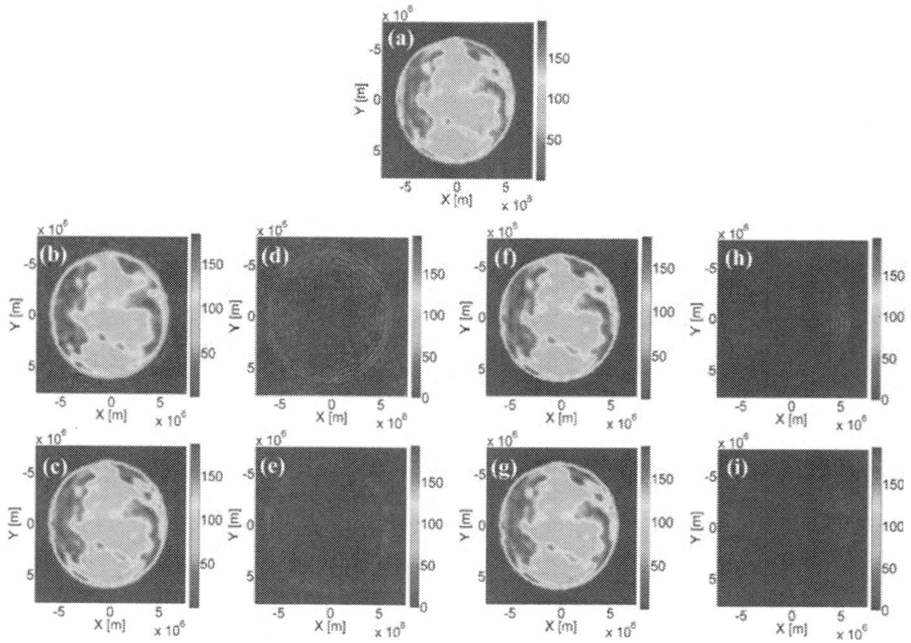

Figure 17. Image reconstruction for the metamaterial-based CRA and GeoSTAR configurations: (a) Original image; reconstruction with pseudo-inverse method for (b) metamaterial-based CRA, and (c) GeoSTAR; error in pseudo-inverse method for (d) metamaterial-based CRA, and (e) GeoSTAR; reconstruction with iterative NESTA method for (f) metamaterial-based CRA, and (g) GeoSTAR; error in NESTA method for (h) metamaterial-based CRA, and (i) GeoSTAR.

analogy between our system and the GeoSTAR system, the frequency range and the largest dimension of the aperture for both configurations are set to be equal. However, the metamaterial-based CRA uses only one-half of the horns required by the GeoSTAR configuration, resulting in less data required for the reconstruction.

Figure 17 shows the original and reconstructed images for the metamaterial-based CRA and GeoSTAR configurations. The error is computed using the Frobenius norm of the difference between the original and the reconstructed images, and it is normalized by the Frobenius norm of the original image. The reconstructed physical temperature using the pseudo-inverse method for the metamaterial-based CRA and the GeoSTAR configuration produces an error of 16.7 and 8.6%, respectively.

The error value of the CS NESTA imaging algorithm for the metamaterial-based CRA and the GeoSTAR configuration is 6.9 and 5.5%, respectively. This shows that the number of receivers are substantially reduced for the metamaterial-based CRA when compared to that of a GeoSTAR system (from 18 to 9), while keeping similar imaging performance.

4. Conclusion

In this chapter, a novel beamforming technique based on CRA for sensing and imaging applications was presented. The CRA uses PEC scatterers and/or MMAs on the surface of the reflector to generate spatial and spectral electromagnetic codes in the imaging domain. The CRA has the ability to increase the sensing capacity of the imaging system, which maximizes the information transfer efficiency. The CRA can reduce the number of feeding elements; therefore, it results in a reduction of the energy budget and the system's complexity. Different examples for active and passive imaging in the mm-wave band were discussed in this chapter. In all examples, the sensing capacity of the CRA was improved when compared to that of the TRA, which ultimately results in a better image reconstruction.

Acknowledgements

This work has been partially funded by the National Science Foundation, CAREER program, Award No. 1653671, and by the Department of Homeland Security, Award No. 2013-ST-061-ED0001. The authors would like to thank Dr. Hipolito Gomez Sousa and Prof. Oscar Rubinos Lopez, from University of Vigo, for their collaboration in the generation of **Figure 5**.

Author details

Ali Molaei[1], Juan Heredia Juesas[1,2] and Jose Angel Martinez Lorenzo[1,2]*

*Address all correspondence to: jmartinez@coe.neu.edu

1 Electrical & Computer Engineering, Northeastern University, Boston, Massachusetts, USA

2 Mechanical & Industrial Engineering, Northeastern University, Boston, Massachusetts, USA

References

[1] Martinez-Lorenzo JA, Heredia-Juesas J, Blackwell W. A single-transceiver compressive reflector antenna for high-sensing-capacity imaging. IEEE Antennas and Wireless Propagation Letters. 2016;**15**:968–971.

[2] Heredia-Juesas J, Blackwell W. Single-transceiver compressive antenna for high-capacity sensing and imaging applications. In: 9th European Conference on Antennas and Propagation (EuCAP); 2015, pp. 1–3.

[3] Heredia-Juesas J, Allan G, Molaei A, Tirado L, Blackwell W, Lorenzo. Consensus-based imaging using ADMM for a compressive reflector. In: IEEE International Symposium on Antennas and Propagation & USNC/URSI National Radio Science Meeting; 2015, pp. 1304–1305.

[4] Molaei A, Heredia-Juesas J, Allan G, Martinez-Lorenzo J. Active imaging using a metamaterial-based compressive reflector antenna. In: IEEE International Symposium on Antennas and Propagation (APSURSI), IEEE; 2016, pp. 1933–1934.

[5] Molaei A, Allan G, Heredia-Juesas J, Blackwell W, Martinez-Lorenzo J. Interferometric sounding using a compressive reflector antenna. In: 10th European Conference on Antennas and Propagation (EuCAP), IEEE; 2016, pp. 1–4.

[6] Gomez-Sousa H, Rubinos-Lopez O, Martinez-Lorenzo JA. Hematologic characterization and 3d imaging of red blood cells using a compressive nano-antenna and ml-fma modelling. In: Antennas and Propagation (EUCAP2016); 2016.

[7] Meana JG, Martinez-Lorenzo JA, Las-Heras F, Rappaport C. Wave scattering by dielectric and lossy materials using the modified equivalent current approximation (MECA). IEEE Transactions on Antennas and Propagation, 2010;**58**(11):3757–3761.

[8] Martinez-Lorenzo JA, Quivira F, Rappaport C. SAR imaging of suicide bombers wearing concealed explosive threats. Progress in Electromagnetics Research. 2012;**125**:255–272.

[9] Candes EJ, Romberg JK, Tao T. Stable signal recovery from incomplete and inaccurate measurements. Communication on Pure and Applied Mathematics, 2006;**59**(8):1207–1223.

[10] Rish I, Grabarnik G. Sparse modeling: theory, algorithms, and Applications. Boca Raton, FL: CRC Press; 2014.

[11] Massa A, Rocca P, Oliveri G. Compressive sensing in electromagnetics—a review. IEEE Antennas and Propagation Magazine, 2015;**57**(1):224–238.

[12] Oliveri G, Anselmi N, Massa A. Compressive sensing imaging of non-sparse 2d scatterers by a total-variation approach within the born approximation. IEEE Transactions on Antennas and Propagation, 2014;**62**(10):5157–5170.

[13] Tse D, Viswanath P. Fundamentals of wireless communication. Cambridge University Press; 2005.

[14] Watts CM, Liu X, Padilla WJ. Metamaterial electromagnetic wave absorbers. Advanced Materials, 2012;**24**(23):98–120.

[15] Landy NI, Sajuyigbe S, Mock JJ, Smith DR, Padilla WJ. Perfect metamaterial absorber. Physical Review Letters, 2008;**100**(20):207402.

[16] Hunt J, Driscoll T, Mrozack A, Lipworth G, Reynolds M, Brady D, Smith DR. Metamaterial apertures for computational imaging. Science. 2013;**339**(6117):310–313.

[17] Chew WC. Waves and fields in inhomogeneous media. New York: IEEE Press; 1995, 522 p.

[18] Ding W, Chen L, Liang CH. Characteristics of electromagnetic wave propagation in biaxial anisotropic left-handed materials. Progress in Electromagnetics Research, 2007; **70**:37–52.

[19] Ahmed SS, Schiess A, Schmidt LP. Near field mm-wave imaging with multistatic sparse 2D-arrays. In: European Radar Conference (EuRAD); 2009.

[20] Bracewell R. The Fourier transform and its applications. McGraw-Hill; 1965.

[21] Ansys HF. v15. Pittsburgh, PA: ANSYS Corporation Software; 2014.

[22] Lambrigtsen B, Tanner A, Gaier T, Kangaslahti P, Brown S. Prototyping GeoSTAR for the PATH mission. In: Proceedings NASA Science Technology Conference; 2007, pp. 19–21.

[23] Nesterov Y. Smooth minimization of non-smooth functions. Mathematical Programming, 2005;**103**(1):127–152.

Detect and Pointing Algorithms Performance for a 2D Adaptive Antenna Array

Tiago Varum, João N. Matos and Pedro Pinho

Additional information is available at the end of the chapter

http://dx.doi.org/10.5772/67524

Abstract

In recent decades, we have witnessed a great progress in wireless communications. The huge amount of data that users expect to access has required an effort to increase the capacity of wireless networks. The main limitation of these communication systems is the increasing interference between channels and multipath fading. Smart antennas technology has emerged, solving some of these problems and improving the performance of wireless networks. This chapter addresses a group of algorithms, directions of arrival (DOA) and beamforming, applied to planar antenna arrays. The algorithms are simulated, and their performance is evaluated in terms of runtime, accuracy and dependence with signal-to-noise ratio (SNR), applied to a smart antenna system.

Keywords: adaptive antenna, direction of arrival, beamforming, planar array

1. Introduction

Adaptive antennas [1–3], often referred to as smart antennas, have a great potential over all the future wireless communications and have been a topic of relevant research and development over the past years.

This type of antennas attracted the attention in the past, mainly by the military branch, although it has a greater progress in recent years, as the result of technological developments, especially in digital processors and new signal-processing techniques. Nowadays, smart antennas are attractive for several areas including military applications, software-defined radio, satellites, mobile communications (especially in base stations), 4G MIMO and the emerging 5G MIMO.

An adaptive antenna consists, in a simple way, of an antenna array and a sophisticated signal processing. It continuously adjusts the feeding of its array elements, and therefore its radiation characteristics, to adapt to the changes in the environment around them, improving the communication.

The dynamic and flexible adjustment of the radiation pattern using a relevant signal-processing capability allows to respond, in real time, to the communication constraints, first by detecting the signal directions of arrival (DOA), and then, according with them, applying beamforming techniques to shape and steer the radiation pattern of the array, by adjusting the locations of the maximums and nulls, to achieve a better reception or transmission.

In this sense, it is possible to recover a weak signal immersed in strong interference environment where signals come from various directions. Furthermore, these antennas enable the reduction of delay spread effects and a greater ability to reduce the impact of multipath fading.

Figure 1 shows the basic block diagram of an adaptive antenna, in the receiving mode, in which some of the different elements that comprise an adaptive antenna are noticeable.

This structure starts by the antenna array and the entire underlying radio frequency (RF) chain, which includes the possible frequency downconversion and analogue-to-digital conversion. Another important component is the DSP (digital signal processor) unit that handles all signal processing of the adaptive antenna.

Using samples of received signal from each array element, DOA algorithms are applied to estimate the arrival directions of the signal or signals that impinge on the antenna array. Then, using an auxiliary intelligence process, the detected signals are divided in terms of signals-of-interest (SOI) and signals-not-of-interest/interferences. Finally, beamforming algorithms are employed using this information (DOA's and SOI's/SNOI's), computing the necessary weights (W_1, W_2, …, W_n) to apply to each array element, obtaining the desired radiation pattern, pointing the antenna to directions of interest signals, and reducing the impact of the interferences.

This chapter is divided into five sections and explores the main concepts presented in the review article [1], from which some tables and figures are reprinted with permission. It starts with an introduction to adaptive antennas, and an adaptive system applied to two-dimensional (2D)

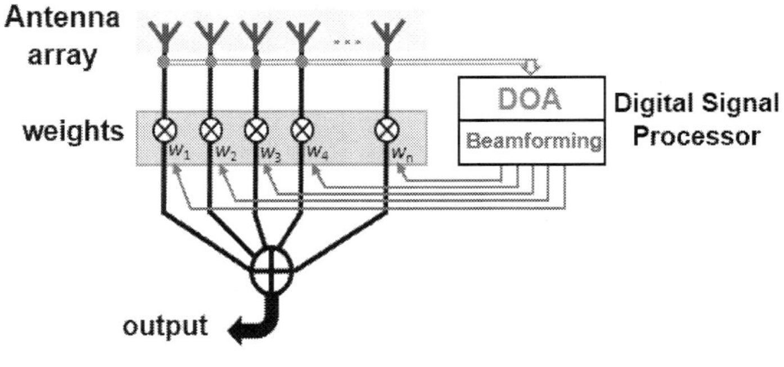

Figure 1. Adaptive antenna block diagram.

http://dx.doi.org/10.5772/67524

arrays is presented in Section 2. In Section 3, two DOA algorithms are presented, which also are tested and compared in terms of important factors as runtime and estimation errors. In Section 4, a group of beamforming algorithms are reported, analysed and applied to a planar 4 × 4 array. Finally, the last section summarizes the main conclusions.

2. Planar antenna array system

The system consists of planar antenna array, as illustrated in **Figure 2**, which is affected by a group of signals including SOI's and SNOI's, resulting in an input signal *x(t)* that is composed by contributions of all these arriving signals, and noise. A sample of this signal *x(t)* is then processed to estimate the angles of arrival of the various signals involved. Afterwards, the respective weights *W* to apply to each element of the array are calculated, to create a proper radiation pattern, pointing it to the direction of interest and reducing the influence of the signals from interfering directions.

Consider a planar antenna array with M × N elements, spaced by d_1 in the rows and d_2 in the columns, in which impinge *J* different signals *s(t)* of frequency with wavelength λ, coming from a direction (θ,ϕ), where θ is the elevation angle and ϕ the azimuth angle. The input signal received by each element *(m,n)* of the array has components from each of *J* arriving signals and noise *n(t)*, and is given by [4, 5]

$$x_{mn} = \sum_{i=1}^{J} s_i(t) e^{j\frac{2\pi}{\lambda}[u_i d_1 (m-1) + v_i d_2 (n-1)]} + n(t) \tag{1}$$

with,

$$u_i = \sin\theta_i, v_i = \cos\theta_i \sin\phi_i, \ m = 1\ldots M, \ n = 1\ldots N, \ i = 1\ldots J$$

The received data *x(t)* and the noise *n(t)* of the whole antenna array can be represented in a vector structure *X(t)* and *N(t)* as follows:

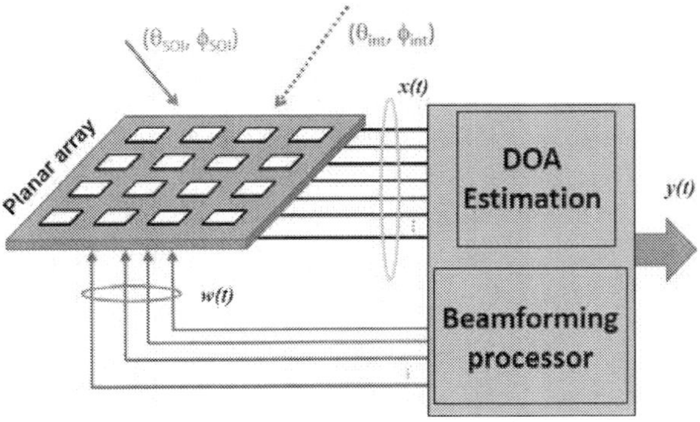

Figure 2. Planar adaptive array system.

$$X(t) = [x_{11}(t)x_{21}(t) \ldots x_{M1}(t)x_{12}(t) \ldots x_{MN}(t)]^T \quad (2)$$

$$N(t) = [n_{11}(t)n_{21}(t) \ldots n_{M1}(t)n_{12}(t) \ldots n_{MN}(t)]^T \quad (3)$$

where T denotes the matrix transpose operation.

The total input signal *X(t)* can be also expressed by the following formula:

$$X(t) = \sum\nolimits_{i=1}^{J} s_i(t)\, A_i + N(t) \quad (4)$$

where A represents the steering matrix of the planar antenna array and is given by Eq. (5). The operation ⊗ denotes the Kronecker product and C_u and C_v are the steering vectors in each direction of the planar array.

The steering vector [6] contains the set of phase delays that a wave will take, relating to each element of the array, and for a planar array this can be represented by Eqs. (6) and (7) [4, 7].

$$A = C_u \otimes C_v \quad (5)$$

$$C_u = \left[1\ e^{j\frac{2\pi d_1}{\lambda}[(2-1)u_i]} \ldots e^{j\frac{2\pi d_1}{\lambda}[(M-1)u_i]}\right]^T \quad (6)$$

$$C_v = \left[1\ e^{j\frac{2\pi d_2}{\lambda}[(2-1)v_i]} \ldots e^{j\frac{2\pi d_2}{\lambda}[(N-1)v_i]}\right]^T \quad (7)$$

The output signal of the adaptive antenna is therefore the product of the received signal *X(t)* by the beamforming weights *W*, estimated through the beamforming algorithms, and applied to each array element, as the following:

$$y(t) = W^H X(t) \quad (8)$$

$$W(t) = [w_{11}(t)w_{21}(t) \ldots w_{M1}(t)w_{12}(t) \ldots w_{MN}(t)]^T \quad (9)$$

In the next sections, a system is implemented and studied using MATLAB [8]. The signal *X(t)*, which consists of the sum of a number of generated signals from several directions (θ,ϕ), being some of these signals SOI and others SNOI, adding up some noise, is created. The locations of the generated incoming signals are totally unknown beyond this step. Then, *X(t)* is sent as a single input parameter for DOA algorithm, which provides as output parameters the estimated number of signals that reach at the array, and their locations (θ,ϕ).

After the locations of all the signals impinging at the antenna are identified, it is necessary to distinguish the signals between the SOI and interferences, and then a beamforming algorithm to determine the necessary weights to apply to each element of the array is applied.

By simulation, using high-frequency structural simulator (HFSS) [9], with the ability to change the amplitude and phase of feeding of a simulated array structure, these sets of estimated weights are applied to the array, and its radiation pattern is analysed.

http://dx.doi.org/10.5772/67524

3. DOA algorithms

The processor that estimates the DOAs of arriving signals to the antenna array is a crucial part of an adaptive antenna, allowing to understand the environment in which the antenna is inserted. By processing the electromagnetic waves that reach an antenna array, it is possible to extract a number of information about them, particularly their arrival directions. This processing is done by using DOA algorithms.

The arriving signals could be divided into SOIs, which are important to steer the antenna towards them, and/or interfering signals (SNOIs), whose impact on the system should be reduced.

There are three main classes of DOA estimation methods referred in the literature, differing mainly in the performance and its computational requirements [10, 11]: the classical, the maximum likelihood and the subspace methods.

The classics are based in the beamforming, in which the central idea is to scan the antenna beam over the space, and the peaks of received power are the DOAs. These methods are theoretically simple but involve a high computational effort and provide a poor performance and a low resolution. The maximum likelihood methods present higher performance than the others (especially with low SNR conditions), because they can take advantage of using better signal and noise models to provide better DOA estimation. However, due to the needs to solve nonlinear multidimensional optimization problems, it increases the required computational load, which makes these methods less popular.

Finally, there are several subspace methods for DOA estimation, which have become popular and widely studied over the last decades due to their good trade-off between the computational complexity and good performance. These methods are based on the eigen decomposition of the estimated covariance matrix of the data received by antenna array, into a signal subspace and a noise subspace. The performance of these methods is essentially limited by the accuracy of distinguishing the signal and the noise subspaces in the presence of noise.

For a planar uniform antenna array, the most applied algorithms are 2D Multiple Signal Classification (MUSIC) and the 2D Estimation of Signal Parameters via Rotational Invariance Techniques (ESPRIT), which are subspace-based methods, and will be subject of a more detailed description below.

3.1. 2D MUSIC algorithm

The Multiple Signal Classification (MUSIC) is perhaps the most popular DOA estimation algorithm, which assumes that the steering vectors of the incoming signals lie in signal subspace, and are orthogonal to the noise subspace. The algorithm search in all possible steering vectors that are orthogonal to the noise subspace of the covariance matrix of the received data (R_{xx}) [11–14].

The MUSIC algorithm, as illustrated in **Figure 3**, uses the received information from each element of the array, and through eigenvalue or singular value decomposition of the R_{xx} matrix, it estimates the noise subspace (U_N).

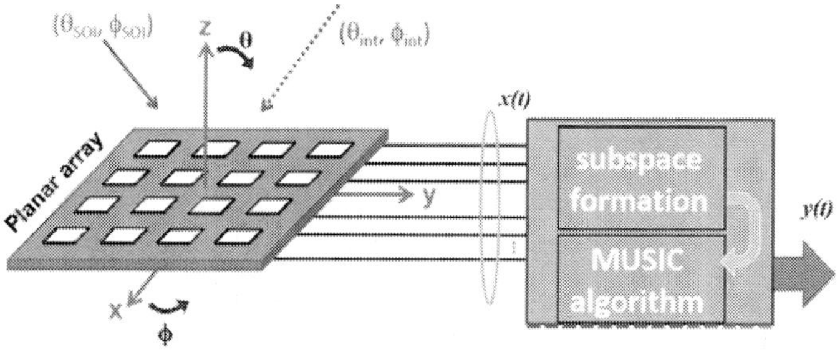

Figure 3. 2D MUSIC DOA estimator.

After the noise subspace U_N is identified, the DOAs are the resulting peaks of the MUSIC spectrum $P_{MUSIC}(\theta,\phi)$, given by Eq. (10) [13]:

$$P_{\text{MUSIC}}(\theta,\phi) = \frac{1}{s^H(\theta,\phi)U_N U_N^H s(\theta,\phi)} \tag{10}$$

where H represents the conjugate transpose matrix (Hermitian).

When a steering vector $s(\theta,\phi)$ is related to one arriving signal, the product of $s^H(\theta,\phi)U_N = 0$, ideally, the function assumes a high value (peak), and therefore (θ,ϕ) is the DOA. There may be several signals from different angles of arrival, creating several peaks in the MUSIC spectrum.

The MUSIC algorithm is simpler to understand and can be applied in all antenna array geometries. However, computationally it requires a lot of resources, since it has to calculate the MUSIC spectrum, Eq. (10), for all the possible steering vectors to estimate the expected peaks.

The estimation error of the MUSIC algorithm is significantly influenced by the angle grid interval in which Eq. (10) is evaluated. In the presence of coherent signals, as in multipath environments, spatial smoothing schemes [15, 16] must be applied to suppress the correlations between the incoming signals.

3.2. 2D ESPRIT algorithm

Estimation of Signal Parameters via Rotational Invariance Techniques (ESPRIT) algorithm is a different subspace-based DOA estimator [17–21]. This algorithm solves some of drawbacks of the MUSIC, in terms of the high computational requirements, and the resulting effects of array calibration errors. The ESPRIT algorithm does not require a high level of calibration in the array since it employs the property of shift invariance of the antenna array. Also, the computational complexity of the ESPRIT is reduced in comparison to MUSIC because it imposes some constraints on array structure.

The ESPRIT algorithm assumes that the separation between equivalent elements in each subarray is fixed, as shown in **Figure 4**, and therefore the array presents a translational invariance.

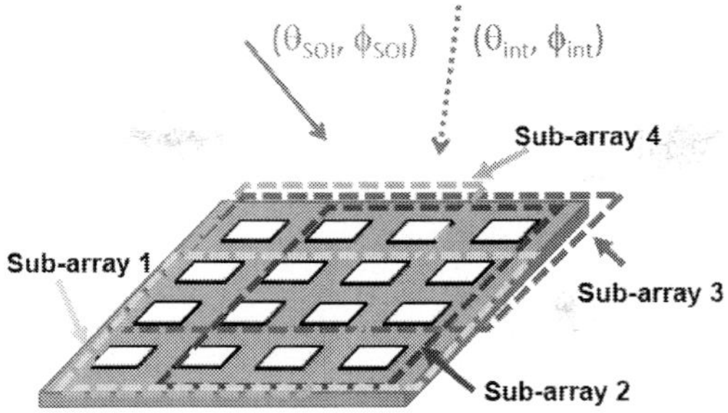

Figure 4. ESPRIT sub-array division with maximum overlap.

This translational invariance leads to a rotational invariance of the signal subspace that will enable to estimate the DOAs.

This algorithm performs three main stages: the signal subspace estimation, the solution of the invariance equation and the DOA estimation.

The algorithm procedure is [17–19] as follows:

Step 1. Signal subspace estimation

Computation of the U_s

Step 2. Solve the invariance equation

$K_{u1}U_sY_u = K_{u2}U_s$

$K_{v1}U_sY_v = K_{v2}U_s$

where K_{u1}, K_{u2}, K_{v1} and K_{v2} represent the two pairs of transformed selection matrices, while Y_u and Y_v are the real-valued matrices.

Step 3. DOA estimation

$\lambda_i\ i = 1\ldots d \rightarrow$ eigen values of $Y_u + \mathrm{j}\ Y_v$

$u_i = 2\ \tan(Re\{\lambda_i\}),$

$v_i = 2\ \tan(Im\{\lambda_i\}),$

$\phi_i = \arg(u_i - \mathrm{j}v_i)\ \theta_i = \sin^{-1}(\| u_i - \mathrm{j}\ v_i \|)$

where θ_i and ϕ_i are the DOA angular information.

3.3. Test and comparison of the DOA algorithms

A system composed by a signal generator (with signals of interest and interferences) followed by a DOA estimator and beamforming was implemented using the MATLAB [8]. The two DOA algorithms, 2D MUSIC and 2D ESPRIT, were accomplished and applied to a planar array configuration.

To test the beamforming results, a planar antenna array of 16 square microstrip elements, arranged in a 4 × 4 planar structure was designed (for a central frequency of 12 GHz), with spacing between elements in both directions of $d = 0.5\lambda$, as shown in **Figure 5**. The array was simulated with HFSS [9] electromagnetic simulator.

Several simulations were performed with good results. As an example, two signals with directions $(\theta,\phi) = (45°, 45°)$ and $(70°, 0°)$ were generated and 'received' by the 4×4 array. With the received signal from each element of the antenna array, given by Eq. (2), the DOAs were estimated using the 2D MUSIC algorithm.

The 2D MUSIC algorithm creates a two-dimensional grid, in the range which the angles vary $\theta \in [0, 90]$ and $\phi \in [0, 360]$, and then evaluates Eq. (10) for each point of the grid.

Figure 6 illustrates the spatial graph of the MUSIC spectrum, resulting in the algorithm, which has peaks in the position of incident signals.

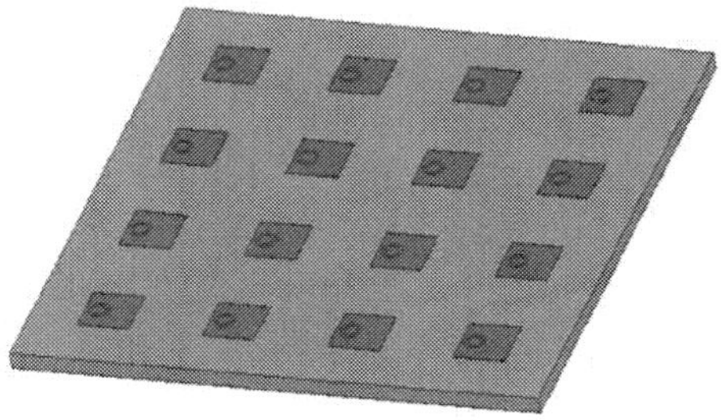

Figure 5. Simulated planar array.

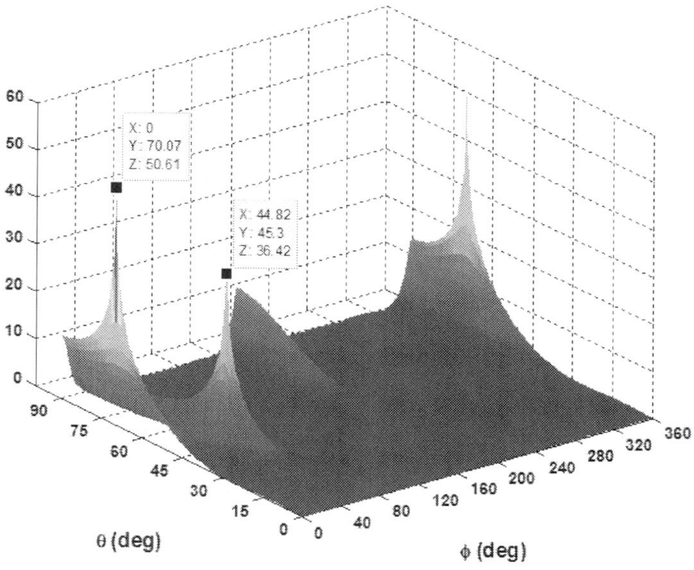

Figure 6. MUSIC spectrum.

According to **Figure 6**, the MUSIC spectrum contains two evidenced peaks. Note that there is another peak but is assumed to be repeated, since 0° and 360° are the same spatial locations.

In order to simplify the definition and extraction of the peaks of the 2D MUSIC spectrum, a function to correctly detect the number of maximum values was implemented. This function provides the points of zero gradient, and its result is shown in **Figure 7**, with the two well-defined peaks.

The output of the 2D MUSIC algorithm is that the incident signals arrive to antenna from (θ,ϕ) (45.3°, 44.82°) and (70.07°, 0°), which are quite close to the initially proposed angles.

Identical signal given to the 2D MUSIC algorithm was provided to the 2D ESPRIT algorithm, also implemented with MATLAB. The output of ESPRIT is just the pair (θ,ϕ) of estimated DOAs, since it does not have a grid to evaluate by a function. The ESPRIT algorithm was preformed and the output result estimates that the signals are arriving from (45.02°, 45.11°) and (69.82°, 0.03°).

The implemented DOA algorithms reveal estimated results very approximate to the original and expected values. These algorithms only receive the signal *X(t)*, and provide the spatial position of each incoming source that compose it.

The analysis of the performance of the 2D DOA algorithms was performed for a large number of experiments. In **Figure 8**, the runtimes of both subspace-based DOA algorithms for $n = 50$ experiences are presented.

The upper part concerns the 2D MUSIC algorithm while the bottom is about the 2D ESPRIT algorithm. In both graphs, the line of the average time of all samples is identified. There is huge

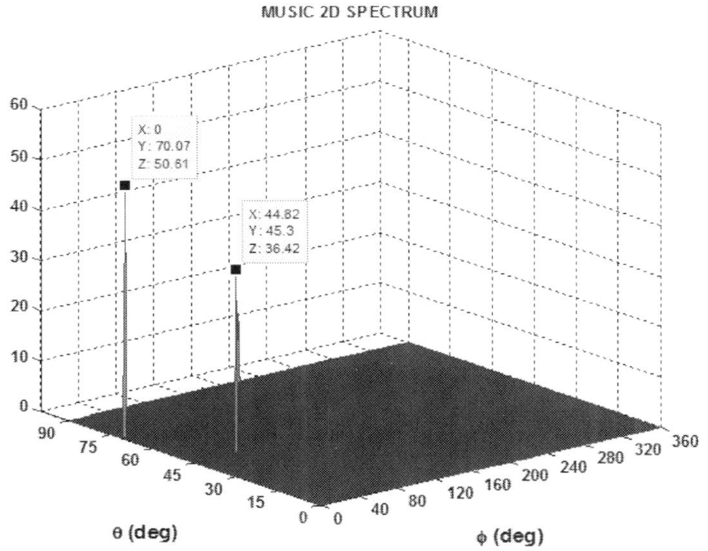

Figure 7. 2D MUSIC spectrum peaks.

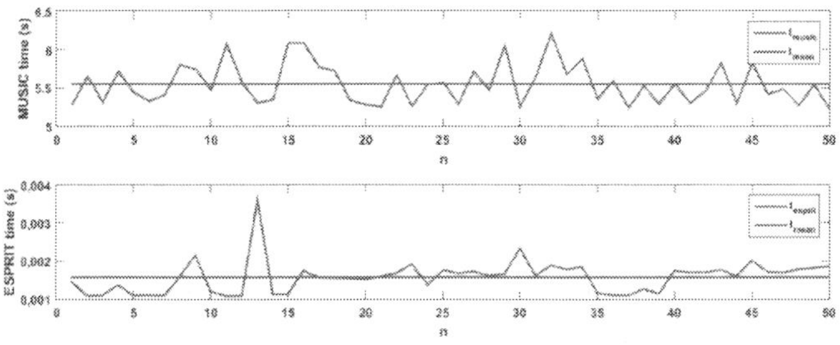

Figure 8. Runtime of DOA algorithms over *n* samples.

difference in the time that MUSIC algorithm takes to perform compared with the ESPRIT. This runtime of the MUSIC is much higher than that of ESPRIT essentially because the MUSIC algorithm needs to calculate the MUSIC function for each possible steering vector.

The 2D MUSIC algorithm takes between 5 and 6 s to estimate the DOAs with an average time of 5.4 s, while 2D ESPRIT, in the majority of samples, varies between 1 and 2 ms, with an average execution time of 1.57 ms.

Another property that can be analysed is the estimation error, between the real location of the incoming waves and the estimated DOAs from both algorithms. **Figure 9** shows the progress of the estimating error over the set of *n* experiences, using the two DOA algorithms. The upper graph is related to the coordinate θ, while the bottom is about the ϕ, and the lines of the average error are further presented.

According to **Figure 9**, it is possible to notice that the 2D MUSIC algorithm has a constant estimation error, with an average value of 0.302° in θ and 0.3052° in ϕ coordinate. The 2D ESPRIT algorithm has average errors extremely lower than 2D MUSIC, about 0.037° in θ and 0.015° in ϕ coordinates.

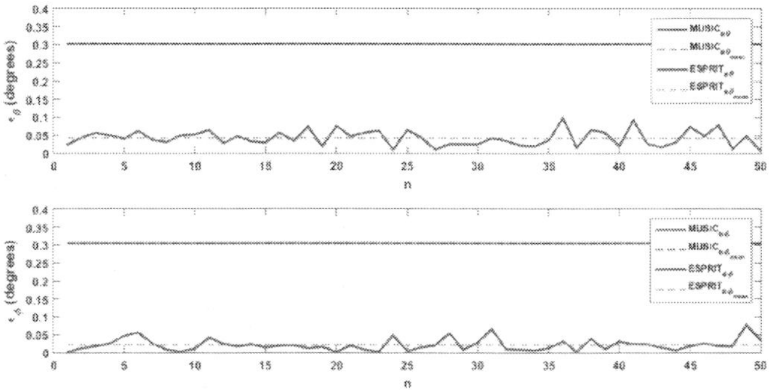

Figure 9. Estimation error in θ and ϕ coordinates for MUSIC and ESPRIT algorithms.

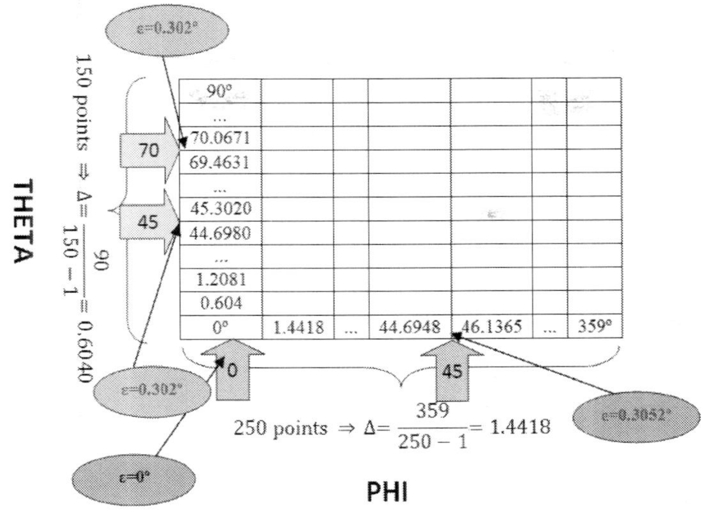

Figure 10. Evaluation grid of 2D MUSIC algorithm.

The 2D ESPRIT errors are mainly due to the mathematical process and the noise of the signal; however, in the 2D MUSIC, the errors are strongly affected by the evaluation interval, as explained in **Figure 10**.

The accuracy of the 2D MUSIC algorithm depends on the number of points of its angle grid in which Eq. (10) is evaluated. More points lead to longer computational times. The number of points of this evaluation grid is then the main issue of MUSIC algorithm, and this value should be an agreement between the required accuracy and computational load allowed.

Figure 10 shows how the MUSIC algorithm of error is consistent with the error estimates of **Figure 9**, and its relationship with the number of grid points, that in these simulations were 150 in θ and 250 in ϕ.

4. Beamforming algorithms

An antenna array, depending on the environment where it is inserted, is usually affected by several electromagnetic signals propagating around, where some are desired to be received, and others, called interferences, are undesirable. Signals of interest and intrusive signals occupy the same frequency range; however, they come from different spatial locations.

In order to improve system performance, the signals of interest must be received in the perfect way, while the impact of the unwanted interferences must be reduced or abolished.

In an adaptive antenna array, after the locations of arriving signals are identified, it is essential to use spatial-filtering techniques, also known as beamforming techniques. These techniques

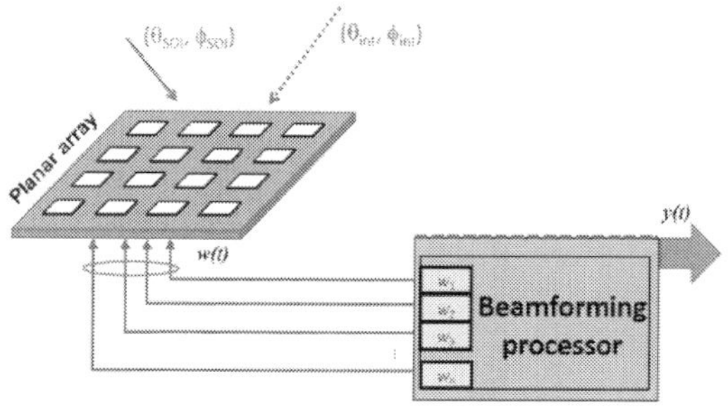

Figure 11. Beamformer system.

deal with the radiation pattern (or beam pattern) of an antenna array, shaping it according to a group of constraints, improving the performance of the communication system.

The beamforming process consists of, as is illustrated in **Figure 11**, changing the relative amplitude and phase of feeding of each element of the array, estimated through beamforming algorithms, adjust the radiation pattern of the array, steering its main lobe and putting nulls in interference directions. There are a number of developed algorithms to estimate the complex weights (amplitude and phase) to apply to an antenna array.

These algorithms are classified as data independent or statistically optimum, according to how the weights are estimated [22, 23]. In the data-independent beamforming algorithms, the weights are estimated to provide a desired response independently of the data received through the antenna, while in the statistically optimum, the weights are estimated according to the statistics of the received data, to optimize its response.

Sometimes, the statistical information of the received data is not available or changes in time; therefore, adaptive algorithms are usually applied to estimate the weights. In this situation, the calculated weights tend to a solution statistically optimum.

4.1. Data-independent algorithms

The data-independent beamformers are characterized by the estimation of its weights to be independent of the received data statistics from the array. One of the most used data-independent algorithms is the classical beamformer [22].

4.1.1. Classical beamformer

The operating principle of this method is similar to the phased array, since it estimates the necessary weights w in order that the maximum of radiation is steered to a desired direction θ_{SOI}. The vector w adjusts the phases to feed each element, in a way that the signals of all the elements are added constructively in a certain direction. However, the main limitation is the impossibility of placing nulls, which at many times is required to eliminate the impact of unwanted signals.

4.2. Statistically optimum algorithms

The statistically optimum beamformers are characterized by the calculation of its weights based on data statistics of received signals. The weights are estimated to steer the radiation pattern in the direction of the SOI while reducing the influence of interfering signals and noises. Some examples of statistically optimum algorithms mentioned in the literature [22] are the multiple side-lobe canceller (MSC), reference signal, SNR maximization and the linearly constrained minimum variance (LCMV).

4.2.1. Multiple side-lobe canceller

The MSC beamformer [22] consists of a main channel and other auxiliary channels. The main idea of this algorithm is to estimate the appropriate weights to apply to the auxiliary channels, in order to cancel or reduce the impact of interference signals in the main channel. In the main channel, a data-independent beamformer can be used to steer the maximum of the array to the desired direction.

Although this method is simple and effective, when the signal of interest is weak relating to interferences, it presents some limitations since the weights are estimated with the absence of the desired signal.

4.2.2. Reference signal

The use of reference signal beamformer [22] requires some knowledge about the desired signal to create a reference signal, and its objective is to minimize the mean square error between the output of the beamformer and the reference signal. The main difficulty of this method is to generate a proper reference signal.

4.2.3. Maximum SNR

The maximum SNR [22] method requires the knowledge of the covariance matrix of the desired signal and of the noise, and the weights are estimated to maximize the SNR.

4.2.4. Linear constrained minimum variance

One of the most important statistically optimum beamformers with higher applicability is the linear constrained minimum variance (LCMV) method [22], and a different approach of its formulation known as generalized side-lobe canceller (GSC).

Most of the times, the desired reference signal is unknown or we do not have enough information about it, being necessary to impose some linear constraints in the weight vector to minimize the variance of beamformer output. The LCMV technique can overcome the main drawbacks of most of the techniques presented before [24].

The principal idea of the LCMV beamformer is that its output is constrained in a way that signals of interest are preserved and the undesired (noise and interfering signals) are minimized.

The LCMV formulation problem is to select the complex weights that are suitable to the multiple linearly independent constraints:

$$\min_w w^H R_x w \text{ Subject to } C^H w = f \tag{11}$$

where w is the vector of weights, R_x the covariance matrix, C is the constraint matrix and f is the response vector.

The solution of the constrained minimization of LCMV problem can be obtained using the method of Lagrange multipliers [22] resulting in

$$w_{\text{opt}} = R_x^{-1} C (C^H R_x C)^{-1} f \tag{12}$$

It is important to observe the dependence of the optimal weight vector of Eq. (12) with the data correlation matrix, and therefore with the statistics of the input signal.

4.2.5. Generalized side-lobe canceller

The generalized side-lobe canceller (GSC) is a different approach to solve the LCMV problem, providing a simple implementation of the beamformer, changing the constrained minimization problem to an unconstrained arrangement [5, 25]. The GSC method, illustrated in **Figure 12**, splits the LCMV problem into two parts, one data independent and other data dependent.

This structure splits the optimum weight vector into two orthogonal components, which are in the range and in the null space of C, in such way that $w = w_0 - B\, w_M$. The beamformer output is $y = w_0{}^H x - w_M{}^H B^H x$.

The vector w_0 is designed to comply with the imposed constraints, consisting in the quiescent part of w. This weight vector is also independent of the input data and represents the non-adaptive component of the LCMV solution. Then, w_0 must satisfy the linear constraints [5]:

$$C^H w_0 = f \Rightarrow w_0 = C (C^H C)^{-1} f \tag{13}$$

The bottom branch of GSC is the data-dependent part, and consists of the blocking matrix B and w_M that will block influence of interfering signals, while minimizing the variance of the output signal y. The blocking matrix B must be orthogonal to the constraint matrix C, so $C^H B = 0$.

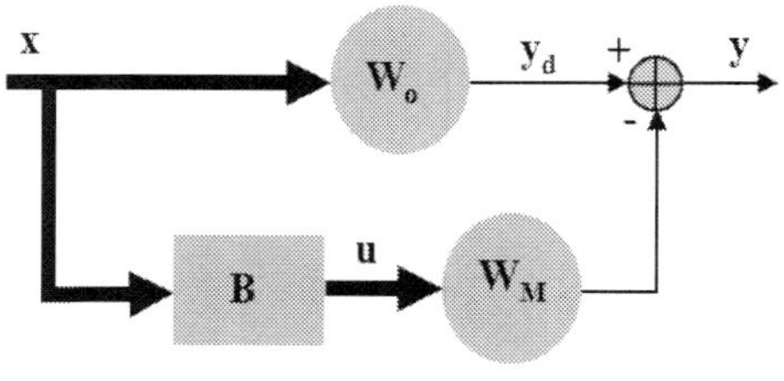

Figure 12. Generalized side-lobe canceller.

http://dx.doi.org/10.5772/67524

The GSC unconstrained problem is

$$\min_{w_M}(w_0 - Bw_M)^H R_x(w_0 - Bw_M) \text{ subject to } C^H w = f \tag{14}$$

and the optimal solution is

$$w_M = (B^H R_x B)^{-1} B^H R_x w_0 \tag{15}$$

This implementation of GSC beamformer enables important benefits, since the w_0 is data-independent beamformer and w_M is an unconstrained beamformer.

4.3. Adaptive algorithms

Adaptive algorithms [22] solve the problem of the statistics of the received data, which changes over the time or may not be available, and affect the statistically optimum beamformers. Examples of adaptive algorithms are the least mean square (LMS), the recursive least squares (RLS) or Frost's algorithm.

4.3.1. Frost's algorithm for LCMV beamforming

Frost's algorithm [23] belongs to the group of LCMV beamformers. Its weights are estimated based on the statistics of the received data, which sometimes are not available or are continually changing, the use of adaptive algorithms being necessary. Frost's algorithm minimizes the mean square error while maintains the specified response to the desired signal. The weight vector starts with an initial value:

$$w = C(C^H C)^{-1} f \tag{16}$$

and at each iteration, the vector is updated on negative-gradient direction with a factor μ, and the weights are defined by

$$w(n+1) = C(C^H C)^{-1} f + P(w(n) - \mu e^*(n)x(n)) \tag{17}$$

$$P = I - C(C^H C)^{-1} C^H \tag{18}$$

4.3.2. Least mean square

Least-mean-square adaptive algorithm estimates the gradient vector from the available data [22, 23, 26, 27]. This method follows an iterative procedure that successively adjusts the weight vector in the direction of the negative of the gradient vector at each iteration, eventually leading to the minimum mean square error:

$$w_M(n) = w_M(n-1) + \mu\, u(n-1)\, y^*(n-1) \tag{19}$$

$$y(n) = y_d(n) - w_M^H(n)u(n) \tag{20}$$

$$0 \leq \mu \leq \frac{1}{\lambda_{max}}$$

where λ_{max} is the largest eigenvalue of the correlation matrix.

The gain μ ($0<\mu<1$) is the parameter that controls the convergence rate of the algorithm. Smaller values lead to slow convergence, but good approximation, while higher values result in a faster convergence and the stability around the minimum value is not ensured. This algorithm is simple since it requires no matrix inversion and by correctly choosing the μ-value, the weight vector tends to an optimum solution.

4.3.3. Recursive least squares

The recursive least squares algorithm has a faster convergence rate than the LMS; however, it involves more complex mathematical operations and requires more computational resources [22, 23].

The RLS problem is

$$\min_{w_{M(k)}} \sum_{n=0}^{N} \lambda^{N-n} |y_d(n) - w_M^H(n)\, u(n)|^2 \tag{21}$$

The RLS calculates the error signal $\alpha(n)$ and is given by

$$\alpha(n) = y_d(n) - w_M^H(n-1)u(n) \tag{22}$$

and then updates the weight vector according to a gain vector $k(n)$

$$w_M(n) = w_M(n-1) + k(n)\alpha^*(n) \tag{23}$$

$$k(n) = \frac{\lambda^{-1}v(n)}{1 + \lambda^{-1}u^H(n)v(n)} \tag{24}$$

The constant λ, $0<\lambda<1$ is called forgetting factor.

$$v(n) = P(n-1)u(n) \tag{25}$$

The matrix $P(n)$ corresponds to the inverse correlation matrix of the input signal which has $P(0)$ as initial value.

$$P(n) = \lambda^{-1}P(n-1) - \lambda^{-1}k(n)v^H(n) \tag{26}$$

$$P(0) = \delta^{-1}I \tag{27}$$

4.4. Test and comparison of the beamforming algorithms

The beamforming weights can be exhibited in the exponential form $w = Ae^{\phi}$, and consist of a set of amplitude and phases, that are applied to each element of an antenna array, to combine its signals in such a way that it produces constructive interference for the desired locals and destructive for undesired.

http://dx.doi.org/10.5772/67524

With the MATLAB, some beamforming algorithms were implemented, and the weights from each algorithm were computed. Then, their performance was analysed when applied to the aforementioned planar antenna array, as presented in **Figure 5**.

The system (DOA and beamforming) simulation was tested using several groups of angles of arrival with excellent results, first by estimating the necessary weights, after, applying it to the simulated planar array to analyse its radiation pattern. In the simulations presented here, it was considered that the signals were coming from signal of interest: $(\theta,\phi)=(45°, 45°)$ and interference signal: $(\theta,\phi)=(70°, 0°)$.

The LVCM algorithm was implemented, using as input parameters the two pairs of angles considered, and with a response vector $f = [1\ 0]^H$ in order to assume the first pair of angles the SOI direction, and the second the SNOI. The output weights, given in terms of amplitude and phase, to apply to the corresponding element of the 4 × 4 array, are presented in **Table 1(a)**. The resultant radiation pattern when these weights are applied is shown in **Figure 13(a)**.

Amplitude ∠ Phase				
	1	**2**	**3**	**4**
(a)	Optimum LCMV			
1	1.0 ∠ 0°	1.0 ∠ −90°	1.0 ∠ −180°	1.0 ∠ 90°
2	1.0 ∠ −90°	1.0 ∠ 180°	1.0 ∠ 90°	1.0 ∠ 0°
3	1.0 ∠ 180°	1.0 ∠ 89°	1.0 ∠ 0°	1.0 ∠ −90°
4	1.0 ∠ 90°	1.0 ∠ 0°	1.0 ∠ −91°	1.0 ∠ −180°
(b)	Adaptive Frost's LCMV			
1	1.0 ∠ 0°	1.0 ∠ −90°	1.0 ∠ −180°	1.0 ∠ 90°
2	1.0 ∠ −90°	1.0 ∠ −180°	1.0 ∠ 90°	1.0 ∠ 0°
3	1.0 ∠ 180°	1.0 ∠ 90°	1.0 ∠ 0°	1.0 ∠ −90°
4	1.0 ∠ 90°	1.0 ∠ 0°	1.0 ∠ −90°	1.0 ∠ −180°
(c)	Adaptive LMS			
1	1.0 ∠ 0°	1.0 ∠ −83°	0.9 ∠ 156°	0.6 ∠ 83°
2	0.7 ∠ −83°	1.0 ∠ 179°	1.0 ∠ 93°	1.2 ∠ 2°
3	0.9 ∠ −178°	0.9 ∠ 61°	0.9 ∠ 23°	0.6 ∠ −78°
4	0.9 ∠ 69°	0.6 ∠ 13°	0.6 ∠ −59°	0.6 ∠ −169°
(d)	Adaptive RLS			
1	1.0 ∠ 0°	0.7 ∠ −104°	0.5 ∠ −161°	0.9 ∠ 102°
2	1.2 ∠ −92°	1.2 ∠ 175°	0.8 ∠ 107°	0.9 ∠ −34°
3	0.8 ∠ −163°	0.3 ∠ 127°	0.8 ∠ −100°	0.8 ∠ −73°
4	0.4 ∠ 101°	0.6 ∠ −4°	0.7 ∠ −114°	0.5 ∠ −128°

Table 1. Estimated weights from beamforming algorithms.

Figure 14 shows the 2D gain distribution of the array as a function of θ and ϕ coordinates, for a better analysis of the 3D radiation patterns of **Figure 13**.

Through **Figure 13(a)** together with **Figure 14(a)**, it is possible to observe the maximum of the radiation pattern pointed to the direction of the SOI (45°, 45°), while in the direction (70°, 0°) there is a low-gain value, which reduces significantly the influence of the interference signal in this direction.

The adaptive algorithms were also implemented because they are a more realistic approach, due to the environment changes.

The adaptive Frost's algorithm was tested applying the considered pair of angles. This algorithm was processed with n = 100 samples of the input signal, iteratively. The output resulting weights are shown in **Table 1(b)**, and the produced radiation pattern is presented

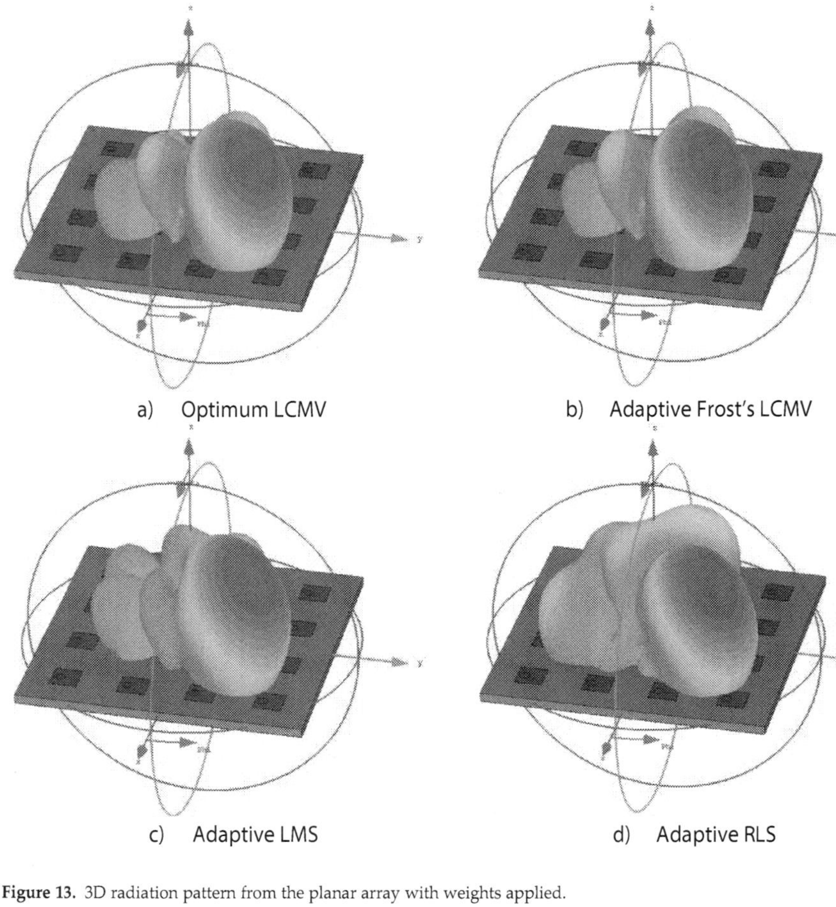

Figure 13. 3D radiation pattern from the planar array with weights applied.

in **Figures 13(b)** and **14(b)**. The array points to the direction of interest and places a null in the interference zone, as desired.

The least-mean-square algorithm was also performed in MATLAB using 100 samples of the received signal *X(t)*. The output of this algorithm is shown in **Table 1(c)**, and the respective radiation pattern is presented in **Figure 13(c)**. Through **Figures 13(c)** and **14(c)**, it is possible to find out that using these weights, the antenna steers its main lobe in the direction of interest (45°, 45°) and places a null in the SNOI zone (70°, 0°).

Finally, the RLS algorithm was implemented and its weights for identical scenario were estimated. The output weights are shown in **Table 1(d)**, leading to the radiation pattern of **Figure 13(d)**. The array directs its maximum of radiation pattern to SOI direction (45°, 45°) as can be seen in **Figure 14(d)**.

The four beamforming algorithms being implemented and tested, a set of 50 consecutive experiments were done, performing a comparative statistical analysis, evaluating the runtime for each iteration. **Figure 15** shows the progress of the execution time of each beamforming algorithm along the *n* experiences. In each sub-figure, a straight line (red) corresponding to the average runtime of all the 50 samples is also presented.

According to **Figure 15**, there are a couple of samples with a more accentuated discrepancy, observed by a couple of peaks in the runtime compared to the average value, property that is common in all beamforming algorithms.

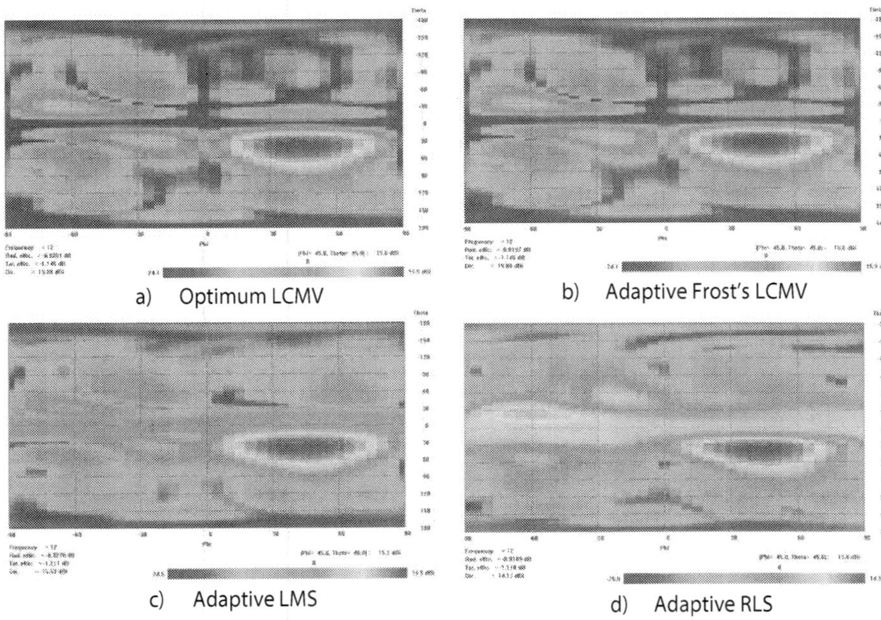

Figure 14. 2D gain distribution of the radiation pattern from the planar array with weights applied.

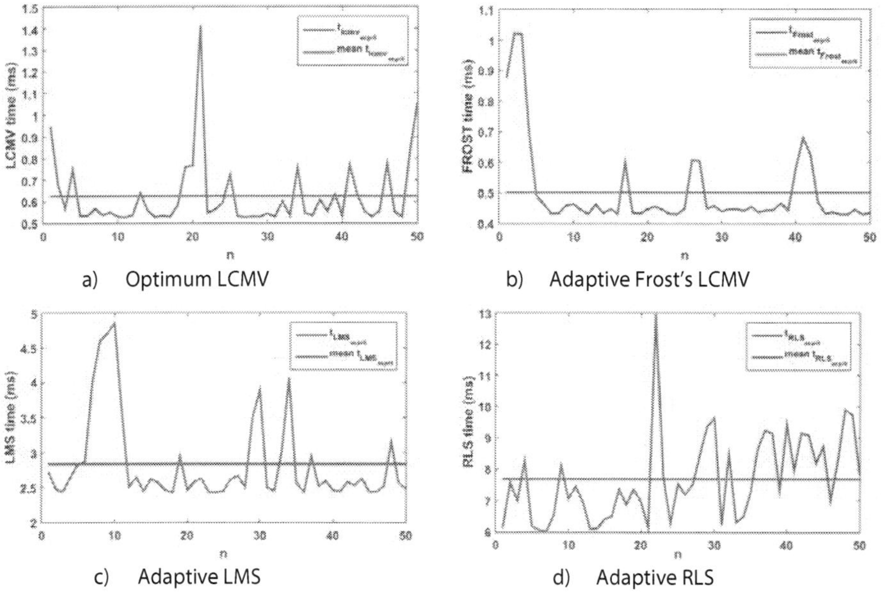

Figure 15. 3D radiation pattern from the planar array with weights applied.

The LCMV algorithm, **Figure 15(a)**, has an average runtime of about 0.62 ms. Concerning the adaptive algorithms, Frost's shown in **Figure 15(b)** present a mean value of 0.5 ms, that even being an adaptive presents a better result than the LCMV (but similar).

The LMS and RLS algorithms, presented in **Figure 15(c)** and, have an average runtime over all the 50 experiences, of about 2.8 and 7.6 ms, with the RLS, as expected, requiring more computational resources than LMS.

After the evaluation of the DOA and beamforming algorithms independently, a comparative analysis about their performance in the system was made, operating together as an adaptive antenna array. The system consists of DOA estimation followed by beamforming weights computation. It was evaluated in terms of total runtime (for all algorithms), the changeability of its results when the noise level modifies, and in terms of estimation errors. The results are presented in **Tables 2–5**, for each beamforming algorithm.

According to **Table 2**, the runtime of LCMV algorithm is similar using both DOA algorithms (MUSIC and ESPRIT), and its variation with SNR is not significantly noted when SNR changes from 10 to 15 dB. However, it almost doubles when SNR changes from 10 to 5 dB. Another important and already referred characteristic is that the MUSIC algorithm is heavier than the ESPRIT in terms of runtime, and while the τ_{MUSIC} does not show a significant variation, the τ_{ESPRIT} tends to reduce with the increase of SNR. In terms of DOA estimation errors, they tend to reduce with the increase of the SNR.

http://dx.doi.org/10.5772/67524

Algorithms		SNR (dB)		
DOA	Performance	5	10	15
MUSIC	τ_{MUSIC} (s)	5.92	6.03	5.34
	τ_{LCMV} (s)	0.0199	0.00089	0.00098
	ε_θ	0.302	0.302	0.302
	ε_ϕ	0.305	0.305	0.305
ESPRIT	τ_{ESPRIT} (s)	0.052	0.038	0.013
	τ_{LCMV} (s)	0.0013	0.0011	0.0009
	ε_θ	0.08	0.04	0.02
	ε_ϕ	0.167	0.141	0.0275

Table 2. Variation of the runtime and estimation error, using LCMV beamforming algorithm.

Algorithms		SNR (dB)		
DOA	Performance	5	10	15
MUSIC	τ_{MUSIC} (s)	5.67	5.15	5.53
	τ_{LCMV} (s)	0.0118	0.0047	0.0007
	ε_θ	0.302	0.302	0.302
	ε_ϕ	0.305	0.305	0.305
ESPRIT	τ_{ESPRIT} (s)	0.015	0.0018	0.0077
	τ_{LCMV} (s)	0.013	0.001	0.00086
	ε_θ	0.140	0.042	0.042
	ε_ϕ	0.0188	0.014	0.010

Table 3. Variation of the runtime and estimation error, using Frost's beamforming algorithm.

Regarding ESPRIT algorithm, a reduction of the estimation errors is observed; however, using the MUSIC algorithm this error maintains constant. This fact is due to the choice of the angle grid for the evaluation of music function P_{MUSIC}, Eq. (10), as was previously referred. Its interval must be an agreement between runtime and estimation error.

Using Frost's algorithm, **Table 3**, its runtime reduces with the SNR. The performance of DOA algorithms maintains with the already described characteristics, in terms of execution time and estimation errors.

The performance using LMS algorithm is presented in **Table 4**. The runtime decreases when the SNR increases, whereas the MUSIC and ESPRIT algorithms maintain with similar characteristics to the preceding cases.

Finally, using the RLS algorithm, **Table 5**, it is possible to observe a high reduction of the runtime when SNR increases from 5 to 10 dB. The error follows the expected behaviour, reducing when SNR increases, using the DOA ESPRIT algorithm, and with the DOA MUSIC algorithm the estimation error holds up constant.

Algorithms		SNR (dB)		
DOA	Performance	5	10	15
MUSIC	τ_{MUSIC} (s)	5.35	5.37	5.46
	τ_{LCMV} (s)	0.020	0.007	0.002
	ε_θ	0.302	0.302	0.302
	ε_ϕ	0.305	0.305	0.305
ESPRIT	τ_{ESPRIT} (s)	0.0095	0.0012	0.0011
	τ_{LCMV} (s)	0.0054	0.0050	0.0046
	ε_θ	0.203	0.042	0.02
	ε_ϕ	0.090	0.045	0.024

Table 4. Variation of the runtime and estimation error, using LMS beamforming algorithm.

Algorithms		SNR (dB)		
DOA	Performance	5	10	15
MUSIC	τ_{MUSIC} (s)	5.82	5.75	5.71
	τ_{LCMV} (s)	0.0163	0.0070	0.0161
	ε_θ	0.302	0.302	0.302
	ε_ϕ	0.305	0.305	0.305
ESPRIT	τ_{ESPRIT} (s)	0.00181	0.00180	0.0015
	τ_{LCMV} (s)	0.0101	0.01	0.01
	ε_θ	0.2294	0.0378	0.0972
	ε_ϕ	0.156	0.031	0.01

Table 5. Variation of the runtime and estimation error, using RLS beamforming algorithm.

5. Conclusions

In this chapter, a set of DOA and beamforming algorithms was studied and implemented. Furthermore, an analysis of each algorithm was carried out, allowing to understand and consolidate the differences between the various algorithms. Regarding the DOA algorithms, both have good estimates in all the simulations performed. It was possible to conclude the reason why the MUSIC algorithm has much larger runtimes and estimation errors; this is due to the grid in which the algorithm is applied. However, it was also concluded that it is a simpler algorithm to implement compared to the ESPRIT. In terms of beamforming algorithms, four algorithms have been implemented and tested and the main notes retained are that all of them present good estimations of weights, and the radiation patterns obtained point to the desired direction. It is also possible to remark the reduction of the runtime of the algorithms with an increase of SNR.

Acknowledgements

The authors are grateful to Fundação para a Ciência e a Tecnologia in the scope of R&D Unit 50008, financed by the applicable financial framework (FCT/MEC through national funds and when applicable co-funded by FEDER-PT2020 partnership agreement under the project UID/EEA/50008/2013), to European Structural and Investment Funds (FEEI) through the Competitiveness and Internationalization Operational Program - COMPETE 2020 and by National Funds through FCT - Foundation for Science and Technology under the project POCI-01-0145- FEDER- 016432, for the partial funding support of this work.

Author details

Tiago Varum[1,2]*, João N. Matos[1,2] and Pedro Pinho[1,3]

*Address all correspondence to: tiago.varum@ua.pt

1 Instituto de Telecomunicações, Aveiro, Portugal

2 Universidade de Aveiro, Aveiro, Portugal

3 Instituto Superior de Engenharia de Lisboa, Lisboa, Portugal

References

[1] T. Varum, J. N. Matos and P. Pinho. Detect and pointing algorithm's performance for a planar smart antenna array: a review. ACES Journal. 2015;**30**(8):824–837.

[2] C. Balanis and P. Ioannides. Introduction to Smart Antennas. Synthesis Lectures on Antennas. Morgan & Claypool Publishers, USA; 2007.

[3] B. Widrow, P. E. Mantey, L. J. Griffiths and B. B. Goode. Adaptive antenna systems. Proceedings of the IEEE. 1967;**55**:2143–2159.

[4] S. J. Yu and J. H. Lee. Design of two-dimensional rectangular array beamformers with partial adaptivity. IEEE Transactions on Antennas and Propagation. 1997;**45**:157–167.

[5] J. H. Lee and Y. H. Lee. Two-dimensional adaptive array beamforming with multiple beam constraints using a generalized sidelobe canceller. IEEE Transactions on Signal Processing. 2005;**53**(9):3517–3529.

[6] L. C. Godara. Applications of antenna arrays to mobile communications. I. Performance improvement, feasibility, and system considerations. Proceedings of the IEEE. 1997;**85**(7):1031–1060.

[7] R. Li, C. Rao, L. Dai and S. Zhao. Adaptive-adaptive beamforming algorithm of planar array based on one-dimensional auxiliary beam. 3rd International Congress on Image and Signal Processing (CISP). 2010;**7**:3300–3303.

[8] MATLAB. Version 7.10.0 (R2010a). Natick, MA: The MathWorks Inc.; 2010.

[9] Ansoft Corporation. User's Guide: High Frequency Structure Simulator, Edition: REV 1.0 Software Version 10.0; 2005.

[10] A. S. Foutz and M. Banavar. Narrowband direction of arrival estimation for antenna arrays. Synthesis Lectures on Antennas: Morgan & Claypool Publishers, USA; 2008.

[11] L. C. Godara. Application of antenna arrays to mobile communications. II. Beamforming. Proceedings of the IEEE. 1997;**85**(8):1195–1245.

[12] R. Schmidt. Multiple emitter location and signal parameter estimation. IEEE Transactions on Antennas and Propagation. 1986;**34**(3):276–280.

[13] S. Sekizawa. Estimation of arrival directions using music algorithm with a planar array. ICUPC '98. IEEE 1998 International Conference on Universal Personal Communications. 1998;**1**(10):555–559.

[14] K. V. Rangarao and S. Venkatanarasimhan. Gold-music: a variation on music to accurately determine peaks of the spectrum. IEEE Transactions on Antennas and Propagation. 2013;**61**(4):2263–2268.

[15] C. C. Yeh, J. H. Lee and Y. M. Chen. Estimating two-dimensional angles of arrival in coherent source environment. IEEE Transactions on Acoustics, Speech, and Signal Processing. 1989;**37**(1):153–155.

[16] Y. M. Chen. On spatial smoothing for two-dimensional direction-of-arrival estimation of coherent signals. IEEE Transactions on Signal Processing. 1997;**45**(7):1689–1696.

[17] Z. Chen, G. Gokeda and Y. Yu. Introduction to Direction-of-Arrival Estimation. Artech House Signal Processing Library: Artech House, USA; 2010.

[18] M. Haardt, M. D. Zoltowski, C. P. Mathews and J. Nossek. 2d Unitary esprit for efficient 2d parameter estimation. ICASSP-95, International Conference on Acoustics, Speech, and Signal Processing. 1995;**3**(3):2096–2099.

[19] M. D. Zoltowski, M. Haardt and C. P. Mathews. Closed-form 2-d angle estimation with rectangular arrays in element space or beamspace via unitary esprit. IEEE Transactions on Signal Processing. 1996;**44**(2):316–328.

[20] Y. Y. Wang and W. W. Chen. A low complexity 2d-doa estimation algorithm using signal decomposition. 4th International in High Speed Intelligent Communication Forum (HSIC). 2012;(5):1–4.

[21] J. Hui and Y. Gang. An improved algorithm of ESPRIT for signal DOA estimation. 2012 International Conference on Industrial Control and Electronics Engineering (ICICEE). 2012;(8):317–320.

[22] B. D. V. Veen and K. M. Buckley. Beamforming: a versatile approach to spatial filtering. IEEE ASSP Magazine. 1988;**5**(4):4–24.

[23] W. Liu and S. Weiss. Wideband Beamforming: Concepts and Techniques. Wireless Communications and Mobile Computing. Wiley, United Kingdom; 2010.

[24] F. Huang, W. Xing Sheng and X. Feng Ma. Efficient parallel adaptive beamforming algorithm for planar array system. 8th International Symposium on Antennas, Propagation and EM Theory. 2008;(11):282–285.

[25] A. M. G. Guerreiro, A. D. D. Neto and F. A. Lisboa. Beamforming applied to an adaptive planar array. Radio and Wireless Conference RAWCON 98. 1998:209–212.

[26] S. Razia, T. Hossain and M. A. Matin. Performance analysis of adaptive beamforming algorithm for smart antenna system. 2012 International Conference on Informatics, Electronics Vision (ICIEV). 2012;(5):946–949.

[27] S. K. Imtiaj, I. S. Misra and R. Biswas. Performance comparison of different adaptive beamforming algorithm in smart antennas. 5th International Conference on Computers and Devices for Communication (CODEC). 2012;(12):1–4.

Chapter 5

Beamformer Based on Quaternion Processes

Jian-wu Tao and Wen-xiu Chang

Additional information is available at the end of the chapter

http://dx.doi.org/10.5772/67859

Abstract

In this chapter, the problem of quaternion beamformer based on linear and widely linear hypercomplex processing is investigated in scenarios, where there exist one signal and one interference that are uncorrelated. First, we introduce brief information about the quaternion algebra and a quaternion model of linear symmetric array with two-component electromagnetic (EM) vector-sensors is presented. Based on array's quaternion model, a quaternion MVDR (QMVDR) beamformer is derived and its performance is analysed. Second, we propose the general expression of a quaternion *semi-widely* linear (QSWL) beamformer and derive its useful implementation and the array's gain expression. Finally, we give the main results of Monte Carlo simulation.

Keywords: quaternion beamforming, hypercomplex processes, widely linear hypercomplex processes, polarization signal processes, EM vector-sensor array

1. Introduction

As an important tool of multidimensional signal processing, the quaternion algebra has been applied to spatio-temporal-polarization beamformer based on an electromagnetic (EM) vector-sensor array [1–5]. The potential advantages of multidimensional signal processing are: (1) the correlation and coupling between each dimension are naturally considered, leading to improved accuracies of signal processing; (2) signals of different geometric nature in different dimensions are being represented as a single signal, leading to reduced complexity of processing approaches. For example, we consider an array consisting of M two-component vector-sensors (If an EM vector-sensor consists of only two components, such as two magnetic loops [6], one electric dipole plus one magnetic loop [7] and two electric dipoles [3], it is referred to a two-component vector-sensor in this chapter.), the output of complex 'long vector' beamformer [8, 9] is $y_c = \mathbf{w}_c^H \boldsymbol{x}_c$ where $\boldsymbol{x}_c = [x_{11}, x_{12}, \ldots x_{M1}, x_{M2}]^T$ is the observed vector of array; $\mathbf{w}_c = [w_{11}, w_{12}, \ldots w_{M1}, w_{M2}]^T$ is a complex weighted vector; the symbol $(.)^H$ denotes the complex conjugation transposition operator. Whereas the output of quaternion-based beamformer [3–5] is $y_h = \mathbf{w}_h^{\Delta} \boldsymbol{x}_h = y_c + j\, y_e$ where

$\boldsymbol{x}_h = [x_{h1}, x_{h2}, \ldots x_{hM}]^T$ is the quaternion-valued, observed vector of array and $x_{hm} = x_{m1} + jx_{m2}$ $(m = 1, \ldots, M)$. $\mathbf{w}_h = [w_{h1}, w_{h2}, \ldots w_{hM}]^T$ is a quaternion-valued, weighted vector and $w_{hm} = w_{m1} + jw_{m2}$ $(m = 1, \ldots, M)$. The symbol $(.)^{\Delta}$ denotes the quaternion conjugation transposition operator and j denotes an imaginary unit of quaternions. Comparing y_h with y_c, we can see that the output of quaternion-based beamformer has one more extra information y_e than the output of complex 'long vector' beamformer. By employing this extra information y_e, we can further improve the performance of beamformer.

In this chapter, our aim is to investigate the beamformer of EM vector-sensor array, based on quaternion processes. First, a QMVDR beamformer and its interference and noise canceller (INC) algorithm are proposed. The output signal to interference-plus-noise ratio (*SINR*) expression of INC algorithm is derived in a scenario where there exist one signal and one interference that are uncorrelated. By analysing the effect of sources parameters on the output *SINR*, the fact is explicitly revealed that even though no separation between the DOA's of the desired signal and interference, the maximum value of output *SINR* can be obtained using the orthogonality between the polarizations of the desired signal and interference. Second, we propose a quaternion *semi-widely* linear beamformer and its useful implementation, i.e., quaternion *semi-widely* linear (QSWL) Generalized sidelobe canceller (GSC). Since the QSWL GSC consists of two-stage beamformers, it has more information than the complex 'long vector' beamformer. The increase in information results in the improvement of the beamformer's performance. By designing the weight vectors of two-stage beamformers, the interference is completely cancelled in the output of QSWL GSC and the desired signal is not distorted.

2. Quaternion algebra and vector-sensor array model

2.1. Quaternion algebra

Quaternion came up in the investigations of constructing multidimensional analogues of the field of complex numbers **C**. The field of quaternion numbers **Q** is also algebra over the field of real numbers **R**. The dimension of this algebra is four, and four basis elements are 1, i, j and k. In field of quaternion numbers **Q**, following multiplication is satisfied

$$i^2 = j^2 = k^2 = -1;\ ij = k = -ji;\ jk = i = -kj;\ ki = j = -ik \tag{1}$$

A quaternion variable $x \in \mathrm{Q}$ has two forms of representation. One form is $x = x_1 + ix_2 + jx_3 + kx_4$, where x_1, x_2, x_3 and x_4 are real coefficients. It is referred to as the **R**-expansion of quaternions. We call x_1 the real/scalar part of x, and it is denoted by $\Re(x)$. $ix_2 + jx_3 + kx_4$ is called the imaginary/vector part of x, and it is denoted by $\Im(x)$. We refer to $x^* = x_1 - ix_2 - jx_3 - kx_4$ as the conjugate of x and $|x| = (x_1^2 + x_2^2 + x_3^2 + x_4^2)^{1/2}$ as the modulus of x. The other form is $x = z_1 + jz_2$, where z_1 and z_2 are complex coefficients. It is called the **C**-expansion of quaternions or *Cayley-Dickson*. $x^* = z_1^* - jz_2$ is the conjugate of x and $|x| = (|z_1|^2 + |z_2|^2)^{1/2}$ is the modulus of x. Several properties of quaternions are discussed in **Table 1** [10–12].

x and *y* denotes two quaternions				
Conjugation	**Norm, noted** $\|.\|$	**Inverse, noted** x^{-1}	**Multiplication**	
$xx^*=x^*x=\|x\|^2$	$\|x\|=0$ if and only if $x=0$	If $x\neq 0$, $x^{-1}=x^*/\|x\|^2$	$a: \in \mathbf{R}$	$ax = xa$
$\|x\|=\|x^*\|$	$\|xy\|=\|yx\|=\|x\|\|y\|$		$c \in \mathbf{C}$	$jc = c^*j$, $jcj^*=c^*$
$(xy)^*=y^*x^*$	$\|x+y\|\leq\|x\|+\|y\|$		$q \in \mathbf{Q}$	$xq \neq qx$

Table 1. Several properties of quaternions.

2.2. Vector-sensor array model

Consider a scenario with one narrowband, completely polarized source, which is travelling in an isotropic and homogeneous medium, impinges on a uniform linear symmetric array from direction (θ, φ). This array consists of $2M$ two-component vector-sensors, which is depicted in **Figure 1**, and the spacing between the adjacent two vector-sensors is assumed to be half wavelength. All the vector-sensors are indexed by $-M,\ldots,-1,1,\ldots,M$ from left to right.

Let the array centre be the phase reference point, two highly complex series $x_{m1}(n)$ and $x_{m2}(n)$ are recorded on first and second components of the mth two-component vector-sensor, respectively. $x_{m1}(n)$ and $x_{m2}(n)$ are given by Ward [10]

$$\begin{bmatrix} x_{m1}(n) \\ x_{m2}(n) \end{bmatrix} = \begin{bmatrix} a_1(\theta,\varphi,\gamma,\eta) \\ a_2(\theta,\varphi,\gamma,\eta) \end{bmatrix} q_m(\theta,\varphi)s(n) \tag{2}$$

where $0\leq \theta \leq \pi$ and $0\leq \varphi < \pi$ denote the incidence source's elevation angle measured from the positive z-axis and the azimuth angle measured from the positive x-axis, respectively. $0\leq \gamma < \pi/2$ represents the auxiliary polarization angle, and $-\pi\leq\eta < \pi$ signifies the polarization phase difference. $a_1(\theta,\varphi,\gamma,\eta)$ and $a_2(\theta,\varphi,\gamma,\eta)$ are the responses on first and second components of two-component vector-sensor, respectively. The two-component vector-sensor consists of one electric dipole plus one magnetic loop co-aligned along the x-axis, where $a_1(\theta,\varphi,\gamma,\eta) = e^{i\eta}\cos\varphi\cos\theta\sin\gamma - \sin\varphi\cos\gamma$ and $a_2(\theta,\varphi,\gamma,\eta) = -e^{i\eta}\sin\varphi\sin\gamma - \cos\varphi\cos\theta\cos\gamma$ [7] or two magnetic loops co-aligned along the x-axis, where $a_1(\theta,\varphi,\gamma,\eta) = -e^{i\eta}\sin\varphi\sin\gamma - \cos\varphi\cos\theta\cos\gamma$ and $a_2(\theta,\varphi,\gamma,\eta) = e^{i\eta}\cos\varphi\sin\gamma - \sin\varphi\cos\theta\cos\gamma$ [6]. $q_m(\theta,\varphi)$ is the spatial phase factor describing wave-field propagation along the array, and $q_m(\theta,\varphi) = q^*_{-m}(\theta,\varphi)$ due to the

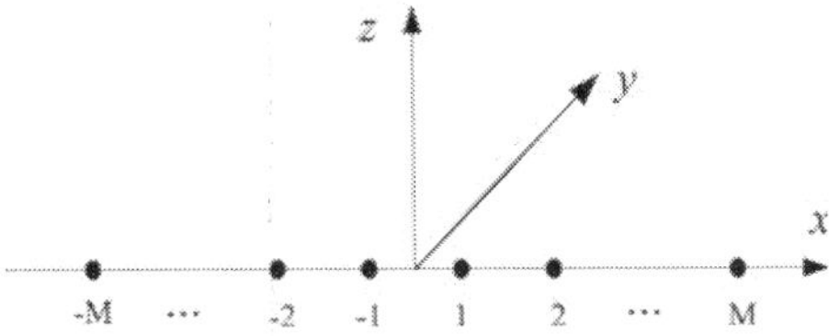

Figure 1. A uniform linear symmetric array.

symmetric structure of array. $s(n)$ is the complex envelope of the waveform, assumed to be a zero-mean, stationary stochastic process.

Using $x_{m1}(n)$ and $x_{m2}(n)$ $(m = -M, \ldots, M)$, a quaternion-valued series $x_m(n)$ can be constructed in the **C**-expansion of quaternions, as the output of the mth two-component vector-sensor:

$$\begin{aligned} x_m(n) &= x_{m1}(n) + jx_{-m2}(n) = q_m(\theta, \varphi)\Big(a_1(\theta, \varphi, \gamma, \eta) + ja_2(\theta, \varphi, \gamma, \eta)\Big)s(n) \\ &= q_m(\theta, \varphi)P(\theta, \varphi, \gamma, \eta)s(n) \end{aligned} \tag{3}$$

where j denotes an imaginary unit of quaternions. $P(\theta, \varphi, \gamma, \eta)$ is the quaternion-valued response on two-component vector-sensor. This transformation maps the complex series $x_{m1}(n)$ on scalar and i imaginary fields of a quaternion, and the complex series $x_{m2}(n)$ is simultaneously mapped to the j and k imaginary fields. When the quaternion-valued additive noise is considered, the quaternion-valued output of the mth two-component vector-sensor is given by

$$x_m(n) = q_m(\theta, \varphi)P(\theta, \varphi, \gamma, \eta)s(n) + n_m(n) \tag{4}$$

where $n_m(n) = n_{m1}(n) + jn_{-m2}(n)$. $n_{m1}(n)$ is the complex-valued additive noises recorded on first component of the mth vector-sensor and $n_{-m2}(n)$ is the complex-valued additive noises recorded on second component of the mth vector-sensor, which are assumed to be zero mean, Gaussian noise with identical covariance σ_n^2. And it is assumed that $n_m(n)$ and $n_n(n)$, where $m \neq n$, are uncorrelated.

3. Quaternion MVDR (QMVDR) beamformer

It is assumed that two uncorrelated, completely polarized plane-waves impinge on an array with $2M$ two-component vector-sensor. One is the desired signal characterized by its arrival angles (θ_s, φ_s) and polarization parameters (γ_s, η_s); the other is the interference characterized by its arrival angles (θ_i, φ_i) and polarization parameters (γ_i, η_i). Assumed interference's DOA and polarization are unknown but signal's DOA and polarization are known or may be priorly estimated from techniques. Thus, the quaternion-valued measurement vector of array can be written as

$$\boldsymbol{x}(n) = [x_{-M}(n), \cdots, x_M(n)]^T = \boldsymbol{v}_s s_s(n) + \boldsymbol{v}_i s_i(n) + \boldsymbol{n}(n) \tag{5}$$

where $\boldsymbol{n}(n) = [n_{-M}(n), \cdots, n_M(n)]^T$ denotes the quaternion-valued additive noise vector. $\boldsymbol{v}_s = \boldsymbol{q}(\theta_s, \varphi_s)P(\theta_s, \varphi_s, \gamma_s, \eta_s), \boldsymbol{v}_i = \boldsymbol{q}(\theta_i, \varphi_i)P(\theta_i, \varphi_i, \gamma_i, \eta_i)$ are the quaternion-valued steering vector associated with the desired signal and the interference, respectively, where $\boldsymbol{q}(\theta_\tau, \varphi_\tau) = [q_{-M}(\theta_\tau, \varphi_\tau), \cdots, q_M(\theta_\tau, \varphi_\tau)]^T$ $(\tau = s, i)$ denotes the spatial phase factor vector of array.

Using the quaternion-valued measurement vector of an array $\boldsymbol{x}(n)$, the output of a beamformer is

$$y(n) = \mathbf{w}^{\Delta}\boldsymbol{x}(n) \tag{6}$$

http://dx.doi.org/10.5772/67859

where $\mathbf{w}$ is the quaternion-valued weight vector and the symbol $(.)^{\Delta}$ denotes the quaternion transposition-conjugation operator. Then, the QMVDR beamformer can be derived by solving the following constrained optimization problem [2]:

$$J(\mathbf{w}) = \min\{\mathbf{w}^{\Delta}\mathbf{R}_x\mathbf{w}\};\, subject\ to\ \mathbf{w}^{\Delta}\boldsymbol{v}_s = 1 \tag{7}$$

where $\mathbf{R}_x = E\{\boldsymbol{x}(n)\boldsymbol{x}^{\Delta}(n)\}$ is the covariance matrix of the measurement vector. By using Lagrange multipliers, the solution of Eq. (7) is obtained, i.e.,

$$J = \mathbf{w}^{\Delta}\mathbf{R}_x\mathbf{w} + \lambda\mathbf{w}^{\Delta}\boldsymbol{v}_s \tag{8}$$

where λ is a real number. Based on the quaternion-valued gradient operator defined in Ref. [11], the following gradients need to be calculated:

$$\frac{\partial J}{\partial \mathbf{w}^{\Delta}} = \mathbf{R}_x\mathbf{w} + \lambda\boldsymbol{v}_s \tag{9}$$

Let Eq. (9) is equal to zero, then

$$\mathbf{w} = -\lambda\mathbf{R}_x^{-1}\boldsymbol{v}_s \tag{10}$$

Since $\mathbf{w}^{\Delta}\boldsymbol{v}_s = -\lambda\boldsymbol{v}_s^{\Delta}\mathbf{R}_x^{-1}\boldsymbol{v}_s = 1$, we have,

$$\lambda = \frac{-1}{\boldsymbol{v}_s^{\Delta}\mathbf{R}_x^{-1}\boldsymbol{v}_s} \tag{11}$$

Substituting Eq. (11) into Eq. (10), the weight vector of the QMVDR beamformer can be written as

$$\mathbf{w} = \frac{\mathbf{R}_x^{-1}\boldsymbol{v}_s}{\boldsymbol{v}_s^{\Delta}\mathbf{R}_x^{-1}\boldsymbol{v}_s} \tag{12}$$

Substituting Eq. (12) into Eq. (6), the quaternion-valued output of the QMVDR beamformer is given by

$$y(n) = s_s(n) + \mathbf{w}^{\Delta}\boldsymbol{v}_i s_i(n) + \mathbf{w}^{\Delta}\boldsymbol{n}(n) \tag{13}$$

where $\mathbf{w}^{\Delta}\boldsymbol{v}_s = 1$. To a linear symmetric array with $2M$ two-component vector-sensors, the signal to interference-plus-noise ratio ($SINR$) of quaternion-valued output $y(n)$ can be written in the simple form (the proof is in Appendix 1 of Ref. [4])

$$SINR_y = \xi_s|P_s|^2(M - \frac{|P_i|^2|\boldsymbol{q}_s^H\boldsymbol{q}_i|^2}{4\xi_i^{-1} + 4M|P_i|^2}) \tag{14}$$

where the input signal-to-noise ratio (SNR) $\xi_s = \frac{\sigma_s^2}{\sigma_n^2}$ and the input interference-to-noise ratio (INR) $\xi_i = \frac{\sigma_i^2}{\sigma_n^2}$.

4. Interference and noise canceller (INC) based on QMVDR

Using the **C**-expansion of quaternions, $y(n)$ can be written as

$$y(n) = y_1(n) + j\, y_2(n) \tag{15}$$

where $y_1(n)$ and $y_2(n)$ are two complex-valued components of $y(n)$, i.e.,

$$y_1(n) = s_s(n) + (\mathbf{w}^{\Delta}\boldsymbol{v}_i)_1 s_i(n) + (\mathbf{w}^{\Delta}\boldsymbol{n}(n))_1;\ y_2(n) = (\mathbf{w}^{\Delta}\boldsymbol{v}_i)_2 s_i(n) + (\mathbf{w}^{\Delta}\boldsymbol{n}(n))_2 \tag{16}$$

$(.)_1$ and $(.)_2$ denote, respectively, first and second complex-valued components of a quaternion. The expansion (16) highlights the fact that $y_2(n)$ does not include the desired signal, but include only the interference and noise. By employing $y_2(n)$, we can partly cancel the interference and noise component in $y_1(n)$. Thus, an INC based on the QMVDR is shown in **Figure 2**.

The INC is a spatio-temporal processing, i.e., first part is a spatio filter and second part is a temporal filter. Then, the output of INC may be written as

$$\begin{aligned} y_s(n) &= y_1(n) - w_c^*\, y_2(n) \\ &= s_s(n) + ((\mathbf{w}^{\Delta}\boldsymbol{v}_i)_1 - w_c^*(\mathbf{w}^{\Delta}\boldsymbol{v}_i)_2) s_i(n) + ((\mathbf{w}^{\Delta}\boldsymbol{n}(n))_1 - w_c^*(\mathbf{w}^{\Delta}\boldsymbol{n}(n))_2) \\ &= s_s(n) + w_i s_i(n) + \varepsilon(n) \end{aligned} \tag{17}$$

where w_c is a complex weight, which can be given by the Wiener-Hoft equation

$$w_c = \frac{r_{y_2 y_1}}{R_{y_2}} \tag{18}$$

where $r_{y_2 y_1} = E\{y_2(n) y_1^*(n)\}$ and $R_{y_2} = E\{y_2(n) y_2^*(n)\}$. Following the proof given in Appendix 2 of Ref. [4], we have

$$w_c = \frac{(\mathbf{w}^{\Delta}\boldsymbol{v}_i)_2 (\mathbf{w}^{\Delta}\boldsymbol{v}_i)_1^* \xi_i}{|(\mathbf{w}^{\Delta}\boldsymbol{v}_i)_2|^2 \xi_i + \|\mathbf{w}\|^2} \tag{19}$$

If ξ_i is very small, w_c is approximately equal to 0. Whereas, if ξ_i is very large, w_c is approximately equal to $\frac{(\mathbf{w}^{\Delta}\boldsymbol{v}_i)_1^*}{(\mathbf{w}^{\Delta}\boldsymbol{v}_i)_2^*}$. In this case, the interference can be cancelled. From Eq. (17), the *SINR* in the complex output $y_s(n)$ is given by

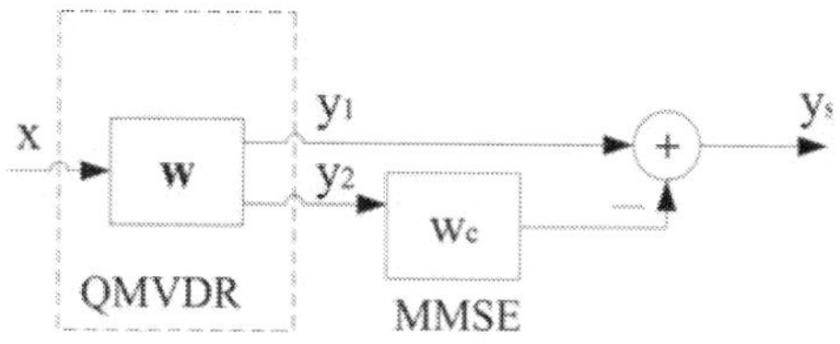

Figure 2. INC based on the QMVDR.

http://dx.doi.org/10.5772/67859

$$SINR_{y_s} = \frac{\xi_s}{\|\mathbf{w}\|^2}\left(1 - \frac{\sigma_i^2|(\mathbf{w}^\Delta \boldsymbol{v}_i)_1|^2}{\sigma_n^2\|\mathbf{w}\|^2 + \sigma_i^2|\mathbf{w}^\Delta \boldsymbol{v}_i|^2}\right) = \kappa(1-\kappa_i) \tag{20}$$

where we define $\kappa = \frac{\xi_s}{\|\mathbf{w}\|^2}$ and $\kappa_i = \frac{\sigma_i^2|(\mathbf{w}^\Delta \boldsymbol{v}_i)_1|^2}{\sigma_n^2\|\mathbf{w}\|^2+\sigma_i^2|\mathbf{w}^\Delta \boldsymbol{v}_i|^2}$. The proof is in Appendix 3 of Ref. [4]. Clearly, $SINR_{y_s}$ increases with an increase in κ but decreases with an increase in κ_i.

Next, we show the effect of sources parameters on κ and κ_i. Following the proof given in Appendix 4 of Ref. [4], we have

$$\kappa = \xi_s|P_s|^2\left(2M - \frac{|P_i|^2|\boldsymbol{q}_s^H\boldsymbol{q}_i|^2}{2\xi_i^{-1} + 2M|P_i|^2}\right)\frac{2M-\mu}{2M-\mu(1+\varepsilon)} = 2SINR_y\beta \tag{21}$$

where

$$\beta = \frac{2M-\mu}{2M-\mu(1+\varepsilon)};\ \mu = \frac{|P_i|^2|\boldsymbol{q}_s^H\boldsymbol{q}_i|^2}{2\xi_i^{-1}+2M|P_i|^2};\ \varepsilon = \frac{\xi_i^{-1}}{\xi_i^{-1}+M|P_i|^2} \tag{22}$$

Obviously, the gain $\beta \geq 1$ because of $0 \leq \varepsilon < 1$. Then, $\kappa \geq 2SINR_y$.

From Eq. (21), κ depends mainly on separation between the DOAs of the desired signal and interference (i.e., $|\boldsymbol{q}_s^H\boldsymbol{q}_i|^2$). The dependencies of κ on $|\boldsymbol{q}_s^H\boldsymbol{q}_i|^2$ are shown in the following consequences:

1. When $\boldsymbol{q}_s = \boldsymbol{q}_i$ (no separation in DOA), $\mu = \frac{2|P_i|^2M^2}{\xi_i^{-1}+M|P_i|^2}$ because of $\boldsymbol{q}_s^H\boldsymbol{q}_i = 2M$. Then, $\beta = 1 + M\xi_i|P_i|^2$ and $\kappa = 2M\xi_i|P_i|^2$. In the case that M is constant, β increases with an increase in ξ_i and $|P_i|^2$, but κ increases with an increase in ξ_s and $|P_s|^2$.

2. When the separation between the DOAs of the desired signal and interference increases, $|\boldsymbol{q}_s^H\boldsymbol{q}_i|$ decreases. This results in the reduction of μ. Then, the both β and κ also reduce. When $\mu = \frac{2|P_i|^2M^2}{2\xi_i^{-1}+M|P_i|^2}$ (i.e., $|\boldsymbol{q}_s^H\boldsymbol{q}_i| = 2M\sqrt{\frac{1+M\xi_i|P_i|^2}{2+M\xi_i|P_i|^2}}$), $\kappa = 2M\xi_s|P_s|^2\left(\frac{4(1+M\xi_i|P_i|^2)}{(2+M\xi_i|P_i|^2)^2}\right)$ reaches to a minimum value. In this case, if $|\boldsymbol{q}_s^H\boldsymbol{q}_i| = \sqrt{2}M$ and $\kappa = 2M\xi_s|P_s|^2$. Along with an increase in ξ_i, the value of $|\boldsymbol{q}_s^H\boldsymbol{q}_i|$, which results in a minimum value of κ, tends to $2M$. Thus, the minimum value of κ tends to 0. Afterwards, κ will increase with a decrease in $|\boldsymbol{q}_s^H\boldsymbol{q}_i|$.

3. When $|\boldsymbol{q}_s^H\boldsymbol{q}_i| = 0$ (i.e., $\mu = 0$), $\beta = 1$ and $\kappa = 2M\xi_s|P_s|^2$. In this case, $\kappa = 2SINR_y$.

In addition, κ depends also on the input INR ξ_i, the array's element number $2M$ and the interference response $|P_i|^2$ and the desired signal response $|P_s|^2$.

In order to illustrate the previous discussions, **Figure 3(a)** and **(b)** displays, respectively, the variations of κ as a function of the desired signal's arrival angles θ_s and φ_s for several values of ξ_i, where $\theta_i = 90°$ $\varphi_i = 60°$; $\varphi_s = 60°$ in **Figure 3(a)** and $\theta_s = 90°$ in **Figure 3(b)**, where a linear symmetric antenna array is used with four (i.e., $M = 2$) two-component vector-sensor.

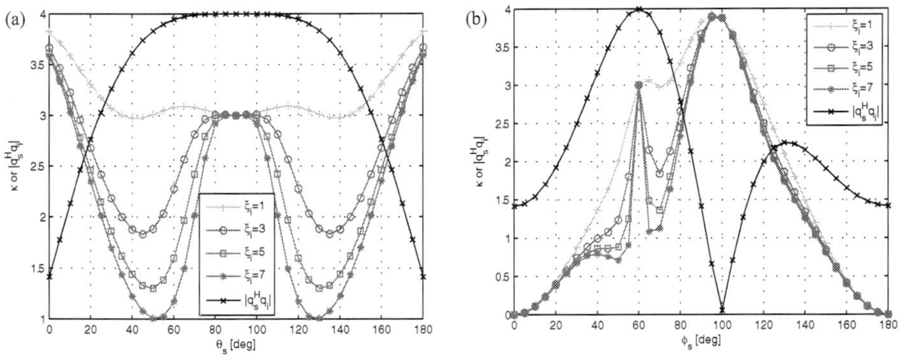

Figure 3. Variations of κ as a function of the desired signal's arrival angles θ_s and φ_s.

Each two-component vector-sensor has one electric dipole and one magnetic loop co-aligned along the x-axis, and four vector-sensors are spaced half a wavelength apart. It is assumed that the desired signal with $\xi_s = 1$ and the interference have the same polarization parameters ($\gamma_s = \gamma_i = 30°, \eta_s = \eta_i = 30°$). Simulation results in **Figure 3** are in agreement with the previous discussions. From **Figure 3(a)**, it is seen that $|\boldsymbol{q}_s^H \boldsymbol{q}_i| \approx 4$ and $\kappa \approx 3$ as $\theta_s \approx 90°$. The cause of this phenomenon is that $|P_s|^2 = \sin^2\varphi_s + \cos^2\varphi_s \cos^2\theta_s$ for the two-component vector-sensor used in this example. When $\theta_s \approx 90°$, $|P_s|^2 \approx 0.75$. In cases of $|\boldsymbol{q}_s^H \boldsymbol{q}_i| = 4$, $\kappa \approx 3$. Along θ_s is away from $90°$, $|\boldsymbol{q}_s^H \boldsymbol{q}_i|$ decreases. This results in the reduction of κ. From **Figure 3(b)**, it is seen that $|\boldsymbol{q}_s^H \boldsymbol{q}_i| \approx 4$ and $\kappa \approx 3$ as $\varphi_s \approx 60°$. This is as same as **Figure 3(a)**. In addition, it is noted that $|\boldsymbol{q}_s^H \boldsymbol{q}_i| \approx 0$ and $\kappa \approx 4$ as $\varphi_s \approx 100°$. The cause of this phenomenon is that $|P_s|^2 \approx 1$ in cases of $\varphi_s \approx 100°$ and $\kappa \approx 4$ in cases of $|\boldsymbol{q}_s^H \boldsymbol{q}_i| = 0$. When $\varphi_s = 0°$ or $\varphi_s = 180°$, $|P_s|^2 = 0$, then, $\kappa = 0$.

Since the output interference-plus-noise power $\mathbf{w}^{\Delta}\mathbf{R}_{in}\mathbf{w} = \sigma_n^2\|\mathbf{w}\|^2 + \sigma_i^2|\mathbf{w}^{\Delta}\boldsymbol{v}_i|^2$, we have

$$\kappa_i = \frac{\sigma_i^2|(\mathbf{w}^{\Delta}\boldsymbol{v}_i)_1|^2}{\sigma_n^2\|\mathbf{w}\|^2 + \sigma_i^2|\mathbf{w}^{\Delta}\boldsymbol{v}_i|^2} = \frac{\sigma_i^2|(\mathbf{w}^{\Delta}\boldsymbol{v}_i)_1|^2}{\mathbf{w}^{\Delta}\mathbf{R}_{in}\mathbf{w}} \tag{23}$$

where $0 \le \kappa_i \le 1$ because of $0 \le |(\mathbf{w}^{\Delta}\boldsymbol{v}_i)_1|^2 \le |\mathbf{w}^{\Delta}\boldsymbol{v}_i|^2$. Following the proof given in Appendix 1 of Ref. [4], κ_i is further written as

$$\kappa_i = \xi_i|P_s|^2\left(M - \frac{|P_i|^2|\boldsymbol{q}_s^H\boldsymbol{q}_i|^2}{4\xi_i^{-1} + 4M|P_i|^2}\right)|(\mathbf{w}^{\Delta}\boldsymbol{v}_i)_1|^2 \tag{24}$$

Obviously, $\kappa_i = 0$ at $|(\mathbf{w}^{\Delta}\boldsymbol{v}_i)_1| = 0$ or $\xi_i = 0$. And κ_i increases with an increase in $|(\mathbf{w}^{\Delta}\boldsymbol{v}_i)_1|^2$. In addition, κ_i depends also on the array's element number $2M$ and the interference response $|P_i|^2$.

Following the proof given in Appendix 5 of Ref. [4], we have

$$(\mathbf{w}^{\Delta}\boldsymbol{v}_i)_1 = \alpha(\boldsymbol{v}_s^{\Delta}\boldsymbol{v}_i)_1 = \alpha|\boldsymbol{q}_s^H\boldsymbol{q}_i|(P_s{}^*P_i)_1 \tag{25}$$

where

$$\alpha = \frac{\xi_i^{-1}}{|P_s|^2}\left(\frac{2}{4M\xi_i^{-1} + |P_i|^2(4M^2 - |\boldsymbol{q}_s^H\boldsymbol{q}_i|^2)}\right) \tag{26}$$

From Eqs. (25) and (26), $(\mathbf{w}^{\Delta}\boldsymbol{v}_i)_1$ depends not only on the separation between the DOAs of the desired signal and interference (i.e., $|\boldsymbol{q}_s^H\boldsymbol{q}_i|$), but also on the difference between the polarizations of the desired signal and interference (i.e., $(P_s{}^*P_i)_1$). The dependencies of $(\mathbf{w}^{\Delta}\boldsymbol{v}_i)_1$ and κ_i on $|\boldsymbol{q}_s^H\boldsymbol{q}_i|$ and $(P_s{}^*P_i)_1$) are shown in the following consequences:

1. When $\boldsymbol{q}_s = \boldsymbol{q}_i$ (i.e., no separation in DOA), $(\mathbf{w}^{\Delta}\boldsymbol{v}_i)_1 = \frac{P_{s1}^*P_{i1}+P_{s2}^*P_{i2}}{|P_s|^2}$ because of $|\boldsymbol{q}_s^H\boldsymbol{q}_i| = 2M$. At the same time, if $P_s = P_i$ (i.e., no difference in polarization), $(\mathbf{w}^{\Delta}\boldsymbol{v}_i)_1 = 1$ because of $P_{s1}^*P_{i1} + P_{s2}^*P_{i2} = |P_s|^2$ So, $\kappa_i = M\xi_i|P_s|^2\left(1 - \frac{M|P_i|^2}{\xi_i^{-1}+M|P_i|^2}\right)$ reaches to a maximum value. Whereas if the polarizations of the desired signal are an orthogonal with that of interference (i.e., $\gamma_s + \gamma_i = \frac{\pi}{2}, \eta_s - \eta_i = \pi$), $(\mathbf{w}^{\Delta}\boldsymbol{v}_i)_1 = 0$ because of $(P_s{}^*P_i)_1 = 0$. So, κ_i reaches to a minimum value.

2. When the separation between the DOAs of the desired signal and interference increases, $|\boldsymbol{q}_s^H\boldsymbol{q}_i|$ decreases. This results in the reduction of $(\mathbf{w}^{\Delta}\boldsymbol{v}_i)_1$. In addition, the increase in the difference between the polarizations of the desired signal and interference also results in the reduction of $(\mathbf{w}^{\Delta}\boldsymbol{v}_i)_1$. Thus, κ_i reduces.

3. When $|\boldsymbol{q}_s^H\boldsymbol{q}_i| = 0$ or $(P_s{}^*P_i)_1 = 0$, $(\mathbf{w}^{\Delta}\boldsymbol{v}_i)_1 = 0$. Thus, $\kappa_i = 0$. In the absence of the interference (i.e., $\xi_i = 0$), $\kappa_i = 0$.

Finally, we analyse the performance of the INC. From Eq. (20) and above analysis, we can obtain the following consequences:

1. When $|\boldsymbol{q}_s^H\boldsymbol{q}_i| = 0$, $\kappa = 2M\xi_s|P_s|^2$ and $\kappa_i = 0$. This implies that the separation between the DOA's of the desired signal and interference reaches to maximum. In this case, we can obtain the maximum value of $SINR_{y_s}$, i.e., $SINR_{y_s\max} = 2M\xi_s|P_s|^2$. Further, $|\boldsymbol{q}_s^H\boldsymbol{q}_i|$ increases with a decrease in the separation in DOA. Thus, $SINR_{y_s}$ will reduce due to the decrease in κ and the increase in κ_i. When $|\boldsymbol{q}_s^H\boldsymbol{q}_i| = 2M\sqrt{\frac{1+M\xi_i|P_i|^2}{2+M\xi_i|P_i|^2}})$, κ reaches to a minimum value. In this case, we can obtain the minimum value of $SINR_{y_s}$ if κ_i reaches to a maximum value.

2. When $(P_s{}^*P_i)_1 = 0$, $\kappa_i = 0$. In this case, we can obtain the maximum value of $SINR_{y_s}$, i.e., $SINR_{y_s\max} = 2M\xi_s|P_s|^2$, if $|\boldsymbol{q}_s^H\boldsymbol{q}_i| = 2M$. This implies that even though no separation between the DOAs of the desired signal and interference, $SINR_{y_s}$ can reach to maximum by using the orthogonally between the polarizations of the desired signal and interference.

Further, $|(P_s{}^*P_i)_1|$ increases with a decrease in the difference in polarizations. Thus, $SINR_{y_s}$ will reduce due to the increase in κ_i.

3. When $\xi_i = 0$ (i.e., in the absence of interference), $\kappa = 2M\xi_s|P_s|^2$ and $\kappa_i = 0$. In this case, we can obtain the maximum value of $SINR_{y_s}$, i.e., $SINR_{y_s\max} = 2M\xi_s|P_s|^2$. Further, $SINR_{y_s}$ decreases with an increase in ξ_i. In the presence of a strong interference (i.e., $\xi_i^{-1} \approx 0$), $SINR_{y_s}$ can be approximated as

$$\mathrm{SINR}_{y_s} \approx \xi_s|P_s|^2\left(2M - \frac{|\boldsymbol{q}_s^H\boldsymbol{q}_i|^2}{2M}\right)\left(1 - \frac{|(\boldsymbol{v}_s{}^{\Delta}\boldsymbol{v}_i)_1|^2}{|\boldsymbol{v}_s{}^{\Delta}\boldsymbol{v}_i|^2}\right) \tag{27}$$

where $\kappa \approx \xi_s|P_s|^2\left(2M - \frac{|\boldsymbol{q}_s^H\boldsymbol{q}_i|^2}{2M}\right)$ and $\kappa_i = \frac{|(\boldsymbol{v}_s{}^{\Delta}\boldsymbol{v}_i)_1|^2}{|\boldsymbol{v}_s{}^{\Delta}\boldsymbol{v}_i|^2}$. Expression (27) highlights the fact that in the presence of a strong interference, $SINR_{y_s} \approx 0$ in the case of no separation between the DOAs of the desired signal and interference (i.e., $\boldsymbol{q}_s = \boldsymbol{q}_i$) or no difference between the polarizations of the desired signal and interference (i.e., $P_s = P_i$). This implies that the INC fails.

4. When the vector-sensor number $2M$ in array increases, κ increases and κ_i decreases. Thus, $SINR_{y_s}$ is an increasing function of $2M$. Since κ increases with an increase in $|P_s|^2$, but κ_i decreases with an increase in $|P_s|^2$, $SINR_{y_s}$ is an increasing function of $|P_s|^2$. Further, $SINR_{y_s}$ is a decreasing function of $|P_i|^2$ because κ decreases with an increase in $|P_i|^2$.

5. The quaternion semi-widely linear (QSWL) beamformer

According to the definition in Ref. [12], the involution of a quaternion x over a pure unit quaternion i is $x^{(i)} = ixi^{-1} = ixi^* = -ixi$ and it represents the reflection of x over the plane spanned by $\{1, i\}$. A quaternion vector $\boldsymbol{x}$ is C^i-proper iff it can be represented by means of two jointly proper complex vectors in the plane spanned by $\{1, i\}$. The augmented covariance matrix of a C^i-proper quaternion vector $\boldsymbol{x}$ can be written as

$$\mathbf{R}_{\overline{\boldsymbol{x}},\overline{\boldsymbol{x}}} = \mathrm{E}\{\overline{\boldsymbol{x}}\,\overline{\boldsymbol{x}}^{\Delta}\} = \begin{bmatrix} \mathbf{R}_{\tilde{\boldsymbol{x}},\tilde{\boldsymbol{x}}} & 0 \\ 0 & \mathbf{R}_{\tilde{\boldsymbol{x}},\tilde{\boldsymbol{x}}}{}^{j} \end{bmatrix} \tag{28}$$

where $\overline{\boldsymbol{x}} = [\boldsymbol{x}^T, \boldsymbol{x}^{(i)T}, \boldsymbol{x}^{(j)T}, \boldsymbol{x}^{(k)T}]^T$ is the augmented quaternion vector; $\tilde{\boldsymbol{x}} = [\boldsymbol{x}^T, \boldsymbol{x}^{(i)T}]^T$ is the semi-augmented quaternion vector and $\mathbf{R}_{\tilde{\boldsymbol{x}},\tilde{\boldsymbol{x}}} = \mathrm{E}\{\tilde{\boldsymbol{x}}\,\tilde{\boldsymbol{x}}^{\Delta}\}$ is the semi-augmented covariance matrix of quaternion vector $\boldsymbol{x}$. In comparison with the semi-augmented covariance matrix $\mathbf{R}_{\tilde{\boldsymbol{x}},\tilde{\boldsymbol{x}}}$, the augmented covariance matrix $\mathbf{R}_{\overline{\boldsymbol{x}},\overline{\boldsymbol{x}}}$ has not more extra information. In other words, the *full-widely* linear processing is equivalent to the *semi-widely* linear processing in handling the C^i-proper quaternion vector. We should not expect that the performance is improved by replacing *semi-widely* linear processing with *full-widely* linear processing.

The most general linear processing is *full-widely* linear processing, which consists in the simultaneous operation on the quaternion vector and its three involutions. Then, a quaternion widely linear beamformer can be written as

$$y(n) = \mathbf{w}^{\Delta}\boldsymbol{x}(n) + \mathbf{G}^{\Delta}\boldsymbol{x}^{(i)}(n) + \mathbf{H}^{\Delta}\boldsymbol{x}^{(j)}(n) + \mathbf{F}^{\Delta}\boldsymbol{x}^{(k)}(n) \tag{29}$$

where **W, G, H** and **F** denote the quaternion-valued weight vectors. $\boldsymbol{x}^{(i)}(n)$, $\boldsymbol{x}^{(j)}(n)$ and $\boldsymbol{x}^{(k)}(n)$ denote the quaternion involution of $\boldsymbol{x}(n)$ over a pure unit imaginary i, j and k, respectively.

The *full-widely* linear processing is optimal processing for the Q-improper quaternion vector. Since the quaternion-valued vector $\boldsymbol{x}(n)$ is C^i-proper vector, however, the optimal processing reduces to *semi-widely* linear processing. Because the *semi-widely* linear processing consists only in the simultaneous operation on the quaternion vector and its involution over i, the general expression of a quaternion *semi-widely* linear (QSWL) beamformer can be written as Ref. [5]

$$y(n) = \mathbf{W}^{\Delta}\boldsymbol{x}(n) + \mathbf{G}^{\Delta}\boldsymbol{x}^{(i)}(n) \tag{30}$$

where $\boldsymbol{x}^{(i)}(n)$ is given by

$$\boldsymbol{x}^{(i)}(n) = -i\boldsymbol{x}(n)i = \boldsymbol{v}_s{}^{(i)}s_s(n) + v_i{}^{(i)}s_i(n) + \boldsymbol{n}^{(i)}(n) \tag{31}$$

Moreover, we can write the quaternion-valued output series $y(n)$ in the following *Cayley-Dickson* representation

$$y(n) = ((\mathbf{W}^{\Delta}\boldsymbol{x}(n))_1 + (\mathbf{G}^{\Delta}\boldsymbol{x}^{(i)}(n))_1) + j((\mathbf{W}^{\Delta}\boldsymbol{x}(n))_2 + (\mathbf{G}^{\Delta}\boldsymbol{x}^{(i)}(n))_2) = y_1(n) + jy_2(n) \tag{32}$$

where y_1 and y_2 denote, respectively, first and second complex-valued components of a quaternion y. Thus, the QSWL beamformer has two complex-valued output series $y_1(n)$ and $y_2(n)$ in the planes spanned by $\{1, i\}$, where $y_1(n) = (\mathbf{W}^{\Delta}\boldsymbol{x}(n))_1 + (\mathbf{G}^{\Delta}\boldsymbol{x}^{(i)}(n))_1)$ and $y_2(n) = (\mathbf{W}^{\Delta}\boldsymbol{x}(n))_2 + (\mathbf{G}^{\Delta}\boldsymbol{x}^{(i)}(n))_2)$. Since the complex 'long vector' beamformers have only one complex-valued output series $y_1(n)$, the QSWL beamformer can obtain more information than the conventional 'long vector' beamformer. The increase of information results in the improvement of QSWL beamformer's performance. In addition, we incorporate both the information on $\boldsymbol{x}(n)$ and $\boldsymbol{x}^{(i)}(n)$, so that the QSWL beamformer with different characteristics may be obtained by designing two weight vectors **W** and **G** under some different criterions.

6. The QSWL generalized sidelobe canceller

In this section, a useful implementation of the QSWL beamformer, i.e., QSWL generalized sidelobe sanceller (GSC), is proposed. The QSWL GSC, which is depicted in **Figure 4**, consists of two-stage beamformers. In first-stage beamformer (weight vector is **W**), we attempt to extract a desired signal without any distortion from observed data. To cancel interferences, we attempt to estimate interferences in second-stage beamformer (weight vector is **G**). By

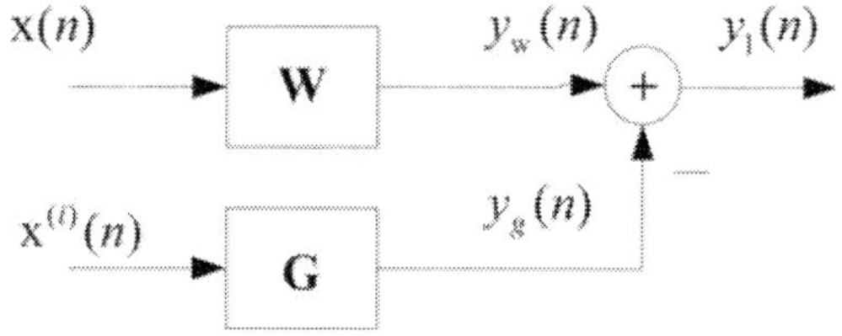

Figure 4. The structure of QSWL GSC.

employing the output of second-stage beamformer to cancel interferences in the output of first-stage beamformer, there is no interference in the output of the QSWL GSC. Compared with the complex 'long vector' beamformers, the advantages of two-stage beamformers are that the main beam can always point to desired signal's direction, even if the separation between the DOAs of the desired signal and interference is less, and the robustness to DOA mismatch is improved.

In the following, we derive the expressions of quaternion-valued weight vectors **W** in first-stage beamformer and **G** in second-stage beamformer. Because the quaternion-valued output $y(n)$ has two complex-valued components in the planes spanned by $\{1, i\}$, i.e., $y_1(n)$ and $y_2(n)$, we define the first complex-valued output component as the output of the QSWL GSC. Thus, the complex-valued output of the QSWL GSC is written as

$$y_{\mathrm{GSC}}(n) = (y(n))_1 = y_w(n) - y_g(n) \tag{33}$$

where $y_w(n)$ is the complex-valued output of the first-stage beamformer, i.e., $y_w(n) = (\mathbf{W}^{\Delta}\boldsymbol{x}(n))_1$; $y_g(n)$ is the complex-valued output of the second-stage beamformer, i.e., $y_g(n) = (\mathbf{G}^{\Delta}\boldsymbol{x}^{(i)}(n))_1$

6.1. The first-stage beamformer

From Eq. (5), we have

$$y_w(n) = (\mathbf{W}^{\Delta}\boldsymbol{v}_s)_1 s_s(n) + (\mathbf{W}^{\Delta}\boldsymbol{v}_i)_1\, s_i(n) + (\mathbf{W}^{\Delta}\boldsymbol{n}(n))_1 \tag{34}$$

In the first-stage beamformer, we attempt to minimize the interference-plus-noise energy in $y_w(n)$, subject to the constraint $(\mathbf{w}^{\Delta}\boldsymbol{v}_s)_1 = 1$.

Since the *Cayley-Dickson* representations of **W**, v_s, v_i and $\boldsymbol{n}(n)$ are, respectively, $\mathbf{W} = \mathbf{W}_1 + j\,\mathbf{W}_2$, $\boldsymbol{v}_s = \boldsymbol{v}_{s1} + j\,\boldsymbol{v}_{s2}$, $\boldsymbol{v}_i = \boldsymbol{v}_{i1} + j\,\boldsymbol{v}_{i2}$ and $\boldsymbol{n}(n) = \boldsymbol{n}_1(n) + j\,\boldsymbol{n}_2(n)$, we have

$$(\mathbf{W}^{\Delta}\boldsymbol{v}_s)_1 = \mathbf{W}_1^H\boldsymbol{v}_{s1} + \mathbf{W}_2^H\boldsymbol{v}_{s2} = \overline{\mathbf{W}}^H\overline{\mathbf{V}}_s \tag{35}$$

$$(\mathbf{W}^{\Delta}\boldsymbol{v}_i)_1 = \mathbf{W}_1^H\boldsymbol{v}_{i1} + \mathbf{W}_2^H\boldsymbol{v}_{i2} = \overline{\mathbf{W}}^H\overline{\mathbf{V}}_i \tag{36}$$

$$(\mathbf{W}^{\Delta}\boldsymbol{n}(n))_1 = \mathbf{W}_1^H\boldsymbol{n}_1(n) + \mathbf{W}_2^H\boldsymbol{n}_2(n) = \overline{\mathbf{W}}^H\overline{\mathbf{N}}(n) \tag{37}$$

where $\overline{\mathbf{W}} = \begin{bmatrix} \mathbf{W}_1 \\ \mathbf{W}_2 \end{bmatrix}$, $\overline{\mathbf{V}}_s = \begin{bmatrix} v_{s1} \\ v_{s2} \end{bmatrix}$, $\overline{\mathbf{V}}_i = \begin{bmatrix} v_{i1} \\ v_{i2} \end{bmatrix}$, $\overline{\mathbf{N}}(n) = \begin{bmatrix} n_1(n) \\ n_2(n) \end{bmatrix}$. Superscript $(.)^H$ denotes the complex conjugate and transpose operator. Thus, Eq. (34) can be rewritten as

$$y_w(n) = \overline{\mathbf{W}}^H \overline{\mathbf{V}}_s s_s(n) + \overline{\mathbf{W}}^H \overline{\mathbf{V}}_i \, s_i(n) + \overline{\mathbf{W}}^H \overline{\mathbf{N}}(n) \tag{38}$$

Then, $\overline{\mathbf{W}}$ can be derived by solving the following constrained optimization problem:

$$J(\overline{\mathbf{W}}) = \min\{\overline{\mathbf{W}}^H \mathbf{R}_{in} \overline{\mathbf{W}}\}; \; \textit{subject to } \overline{\mathbf{W}}^H \overline{\mathbf{V}}_s = 1 \tag{39}$$

where

$$\mathbf{R}_{in} = \begin{bmatrix} \mathrm{E}\{(\boldsymbol{x}_{in}(n))_1 \, (\boldsymbol{x}_{in}(n))_1^H\} & \mathrm{E}\{(\boldsymbol{x}_{in}(n))_1 \, (\boldsymbol{x}_{in}(n))_2^H\} \\ \mathrm{E}\{(\boldsymbol{x}_{in}(n))_2 \, (\boldsymbol{x}_{in}(n))_1^H\} & \mathrm{E}\{(\boldsymbol{x}_{in}(n))_2 \, (\boldsymbol{x}_{in}(n))_2^H\} \end{bmatrix} \tag{40}$$

is the covariance matrix and $\boldsymbol{x}_{in}(n) = \boldsymbol{v}_i s_i(n) + \boldsymbol{n}(n)$ is the measurement vector of array in the absence of the desired signal. The solution of this constrained optimization problem is obtained by using Lagrange multipliers, that is

$$\overline{\mathbf{W}} = \frac{\mathbf{R}_{in}^{-1} \, \overline{\mathbf{V}}_s}{\overline{\mathbf{V}}_s^H \mathbf{R}_{in}^{-1} \, \overline{\mathbf{V}}_s} \tag{41}$$

If the interferences are uncorrelated with the additive noise, $\overline{\mathbf{W}}$ can be written in the simple form (the proof is in Appendix A of Ref. [5])

$$\overline{\mathbf{W}} = \frac{\varepsilon \overline{\mathbf{V}}_s - (P_i{}^{\Delta} P_s)_1 \boldsymbol{q}_i^H \boldsymbol{q}_s \overline{\mathbf{V}}_i}{\mu} \tag{42}$$

where

$$\mu = 2M|P_s|^2 \varepsilon - |(P_i{}^{\Delta} P_s)_1|^2 |\boldsymbol{q}_i^H \boldsymbol{q}_s|^2 \; ; \; \varepsilon = \xi_i^{-1} + 2M|P_i|^2 \tag{43}$$

where ξ_i denotes the input interference-to-noise ratio (INR). Moreover, the quaternion-valued optimal weight vector $\mathbf{W}_o$ may be given by

$$\mathbf{W}_o = J_1 \, \overline{\mathbf{W}} + j \, J_2 \, \overline{\mathbf{W}} \tag{44}$$

where $J_1 = [\mathbf{I}_{2M\times 2M}, \; \mathbf{0}_{2M\times 2M}]$ and $J_2 = [\mathbf{0}_{2M\times 2M}, \; \mathbf{I}_{2M\times 2M}]$ are two selection matrices. It is noted that in some applications, such as Radar, $\mathbf{R}_{in}$ may be estimated in intervals of no transmitted signal. But, $\mathbf{R}_{in}$ is not obtained in other applications, such as Communications. In these applications, we may replace $\mathbf{R}_{in}$ by $\mathbf{R}_x$, where

$$\mathbf{R}_x = \begin{bmatrix} \mathrm{E}\{(\boldsymbol{x}(n))_1 \, (\boldsymbol{x}(n))_1^H\} & \mathrm{E}\{(\boldsymbol{x}(n))_1 \, (\boldsymbol{x}(n))_2^H\} \\ \mathrm{E}\{(\boldsymbol{x}(n))_2 \, (\boldsymbol{x}(n))_1^H\} & \mathrm{E}\{(\boldsymbol{x}(n))_2 \, (\boldsymbol{x}(n))_2^H\} \end{bmatrix} \tag{45}$$

is the covariance matrix and $\mathbf{x}(n)$. When the distortionless constraint is perfectly matched with the desired signal, the weight vector $\mathbf{W}_o$ is identical in both $\mathbf{R}_{in}$ and $\mathbf{R}_x$.

By using the optimal weight vector $\mathbf{W}_o$, the complex output of first-stage beamformer can be given by

$$y_w(n) = s_s(n) + (\mathbf{W}_o{}^{\Delta}\boldsymbol{v}_i)_1\, s_i(n) + (\mathbf{W}_o{}^{\Delta}\boldsymbol{n}(n))_1 \tag{46}$$

6.2. The second-stage beamformer

From Eq. (31), we have

$$y_g(n) = (\mathbf{G}^{\Delta}\boldsymbol{v}_s^{(i)})_1 S_s(n) + (\mathbf{G}^{\Delta}\boldsymbol{v}_i^{(i)})_1\, S_i(n) + (\mathbf{G}^{\Delta}\boldsymbol{n}^{(i)}(n))_1 \tag{47}$$

In the second-stage beamformer, we attempt to minimize the noise energy in $y_g(n)$, subject to the constraints $(\mathbf{G}^{\Delta}\boldsymbol{v}_S^{(i)})_1 = 0$ and $(\mathbf{G}^{\Delta}\boldsymbol{v}_i^{(i)})_1 = (\mathbf{W}_o^{\Delta}\boldsymbol{v}_i)_1$. In the following, two schemes are presented to implement this aim.

6.2.1 The Scheme 1, i.e., combined QPMC and MVDR

Let $= \mathbf{w}_{qs}\mathbf{w}_{MV}$, where $\mathbf{w}_{qs}$ is a quaternion-valued diagonal weight matrix and $\mathbf{w}_{MV}$ is a complex weight vector. In this scheme, the first is to achieve the constraint $(\mathbf{G}^{\Delta}\boldsymbol{v}_s^{(i)})_1 = 0$ by designing $\mathbf{w}_{qs}$, which is referred to quaternion polarization matched cancellation (QPMC); the second is to minimize the noise energy in $y_g(n)$ subject to the constraint $(\mathbf{G}^{\Delta}\boldsymbol{v}_i^{(i)})_1 = (\mathbf{W}_o^{\Delta}\boldsymbol{v}_i)_1$ by designing $\mathbf{w}_{MV}$, which is referred to MVDR.

Let $\mathbf{w}_{qs} = \text{diag}\,\{\mathrm{w}_{qs}(-M),\ \cdots, \mathrm{w}_{qs}(M)\}$; then we have

$$\mathbf{G}^{\Delta}\boldsymbol{v}_s^{(i)} = \mathbf{w}_{MV}{}^H\mathbf{w}_{qs}^{\Delta}\boldsymbol{v}_s^{(i)} = \mathbf{w}_{MV}{}^H\begin{bmatrix}\mathrm{w}_{qs}{}^*(-M)q_{-M}(\theta_s,\ \varphi_s)P_s^{(i)}\\ \vdots \\ \mathrm{w}_{qs}{}^*(M)q_M(\theta_s,\ \varphi_s)P_s^{(i)}\end{bmatrix} \tag{48}$$

where superscript (.)* denotes the quaternion conjugate operator. From Eq. (48) and the constraint $(\mathbf{G}^{\Delta}\boldsymbol{v}_s^{(i)})_1 = 0$, we have the constraint $(\mathrm{w}_{qs}{}^*(m)q_m(\theta_s,\ \varphi_s)P_s^{(i)})_1 = 0$, where $m = \{-M,\ \cdots, M\}$. When $\mathrm{w}_{qs}(m) = q_m(\theta_s,\ \varphi_s)\ (a_{s2}^* + j\, a_{s1}^*)$, this constraint is satisfied. Thus, we can obtain

$$\mathbf{w}_{qs} = \text{diag}\,\{\boldsymbol{q}_s\}\ (a_{s2}^* + j\, a_{s1}^*) \tag{49}$$

where $\text{diag}\,\{\boldsymbol{q}_s\} = \text{diag}\,\{q_{-M}(\theta_s,\ \varphi_s), \cdots, q_M(\theta_s,\ \varphi_s)\}$. In the constraint $(\mathbf{G}^{\Delta}\boldsymbol{v}_s^{(i)})_1 = 0$, we insert $\mathbf{G} = \mathbf{w}_{qs}\mathbf{w}_{MV}$ into (47). Thus, $y_g(n)$ can be rewritten as

$$y_g(n) = \mathbf{w}_{MV}{}^H(\mathbf{w}_{qs}^{\Delta}\boldsymbol{v}_i^{(i)})_1\ s_i(n) + \mathbf{w}_{MV}{}^H(\mathbf{w}_{qs}^{\Delta}\boldsymbol{n}^{(i)}(n))_1 \tag{50}$$

Then, $\mathbf{w}_{MV}$ can be derived by solving the following constrained optimization problem:

$$J(\mathbf{w}_{MV}) = \min\{\mathbf{w}_{MV}{}^H \mathbf{R}_{qs} \mathbf{w}_{MV}\};\ \textit{subject to}\ \mathbf{w}_{MV}{}^H\ \tilde{\mathbf{V}}_i = \overline{\mathbf{W}}^H \overline{\mathbf{V}}_i \tag{51}$$

where $\mathbf{R}_{qs} = \mathrm{E}\{(\mathbf{w}_{qs}^{\Delta} \boldsymbol{x}^{(i)}(n))_1 \left(\mathbf{w}_{qs}^{\Delta} \boldsymbol{x}^{(i)}(n)\right)_1^H\}$ is the covariance matrix and $\tilde{\mathbf{V}}_i = (\mathbf{w}_{qs}^{\Delta} \boldsymbol{v}_i^{(i)})_1$. The solution of this constrained optimization problem is obtained by using Lagrange multipliers, i.e.,

$$\mathbf{w}_{MV} = \frac{\mathbf{R}_{qs}^{-1}\ \tilde{\mathbf{V}}_i}{\tilde{\mathbf{V}}_i{}^H \mathbf{R}_{qs}^{-1} \tilde{\mathbf{V}}_i} \overline{\mathbf{V}}_i{}^H \overline{\mathbf{W}} \tag{52}$$

If the desired signal and interference are uncorrelated with the additive noise, $\mathbf{w}_{MV}$ can be written in the simple form (the proof is in Appendix B of Ref. [5])

$$\mathbf{w}_{MV} = \frac{g_1}{\kappa} \tilde{\mathbf{V}}_i \tag{53}$$

where

$$\kappa = \tilde{\mathbf{V}}_i^H \tilde{\mathbf{V}}_i = 2M(|a_{s2}|^2 |a_{i1}|^2 + |a_{s1}|^2 |a_{i2}|^2) - 2\mathcal{R}(a_{s1} a_{s2}^* a_{i2} a_{i1}^* (\boldsymbol{q}_i{}^2)^H \boldsymbol{q}_s{}^2) \tag{54}$$

$$g_1 = (\overline{\mathbf{W}}^H \overline{\mathbf{V}}_i)^H = \frac{\xi_i^{-1} (P_i{}^{\Delta} P_s)_1 \boldsymbol{q}_i^H \boldsymbol{q}_s}{\mu} \tag{55}$$

where $\mathcal{R}$ (.) denotes the real part of a complex number. μ is given by Eq. (43).

6.2.2. The scheme 2, i.e. Linearly constrained minimum variance (LCMV) beamformer

In this scheme, we employ the LCMV beamformer as the second-stage beamformer. Since the *Cayley-Dickson* representations of $\mathbf{G}$, $\boldsymbol{v}_s^{(i)}$, $\boldsymbol{v}_i^{(i)}$ and $\boldsymbol{n}^{(i)}(n)$ are, respectively, $\mathbf{G} = \mathbf{G}_1 + j\ \mathbf{G}_2$, $\boldsymbol{v}_s^{(i)} = \boldsymbol{v}_{s1} - j\boldsymbol{v}_{s2}$, $\boldsymbol{v}_i^{(i)} = \boldsymbol{v}_{i1} - j\ \boldsymbol{v}_{i2}$ and $\boldsymbol{n}^{(i)}(n) = \boldsymbol{n}_1(n) - j\ \boldsymbol{n}_2(n)$, we have

$$(\mathbf{G}^{\Delta} \boldsymbol{v}_s^{(i)})_1 = \mathbf{G}_1^H \boldsymbol{v}_{s1} - \mathbf{G}_2^H \boldsymbol{v}_{s2} = \overline{\mathbf{G}}^H \overline{\mathbf{V}}_s^{(i)} \tag{56}$$

$$(\mathbf{G}^{\Delta} \boldsymbol{v}_i^{(i)})_1 = \mathbf{G}_1^H \boldsymbol{v}_{i1} - \mathbf{G}_2^H \boldsymbol{v}_{i2} = \overline{\mathbf{G}}^H \overline{\mathbf{V}}_i^{(i)} \tag{57}$$

$$(\mathbf{G}^{\Delta} \boldsymbol{n}^{(i)}(n))_1 = \mathbf{G}_1^H \boldsymbol{n}_1(n) - \mathbf{G}_2^H \boldsymbol{n}_2(n) = \overline{\mathbf{G}}^H \overline{\mathbf{N}}^{(i)}(n) \tag{58}$$

where $\overline{\mathbf{G}} = \begin{bmatrix} \mathbf{G}_1 \\ \mathbf{G}_2 \end{bmatrix}; \overline{\mathbf{V}}_s^{(i)} = \begin{bmatrix} \boldsymbol{v}_{s1} \\ -\boldsymbol{v}_{s2} \end{bmatrix};\ \overline{\mathbf{V}}_i^{(i)} = \begin{bmatrix} \boldsymbol{v}_{i1} \\ -\boldsymbol{v}_{i2} \end{bmatrix};\ \overline{\mathbf{N}}^{(i)}(n) = \begin{bmatrix} \boldsymbol{n}_1(n) \\ -\boldsymbol{n}_2(n) \end{bmatrix}$

Thus, Eq. (47) can be rewritten as

$$y_g(n) = \overline{\mathbf{G}}^H \overline{\mathbf{V}}_s^{(i)} s_s(n) + \overline{\mathbf{G}}^H \overline{\mathbf{V}}_i^{(i)}\ s_i(n) + \overline{\mathbf{G}}^H \overline{\mathbf{N}}^{(i)}(n) \tag{59}$$

Then, $\overline{\mathbf{G}}$ can be derived by solving the following constrained optimization problem:

$$J(\overline{\mathbf{G}}) = \min\{\overline{\mathbf{G}}^H \mathbf{R}_{inG} \overline{\mathbf{G}}\};\ subject\ to\ \overline{\mathbf{G}}^H \mathbf{C} = \mathbf{g}^H \tag{60}$$

where

$$\mathbf{R}_{inG} = \begin{bmatrix} \mathrm{E}\{(\boldsymbol{x}_{in}^{(i)}(n))_1\,(\boldsymbol{x}_{in}^{(i)}(n))_1^H\} & \mathrm{E}\{(\boldsymbol{x}_{in}^{(i)}(n))_1\,(\boldsymbol{x}_{in}^{(i)}(n))_2^H\} \\ \mathrm{E}\{(\boldsymbol{x}_{in}^{(i)}(n))_2\,(\boldsymbol{x}_{in}^{(i)}(n))_1^H\} & \mathrm{E}\{(\boldsymbol{x}_{in}^{(i)}(n))_2\,(\boldsymbol{x}_{in}^{(i)}(n))_2^H\} \end{bmatrix} \tag{61}$$

is the covariance matrix and $\boldsymbol{x}_{in}^{(i)}(n) = \boldsymbol{v}_i^{(i)} s_i(n) + \boldsymbol{n}^{(i)}(n)$ is the quaternion involution of $\boldsymbol{x}_{in}(n)$. $\mathbf{C} = [\overline{\mathbf{V}}_i^{(i)}, \overline{\mathbf{V}}_s^{(i)}]$ and $\mathbf{g}^H = [g_1^H,\ 0]$, where g_1 is given by Eq. (55). The solution of Eq. (60) is given by Van Trees [13]

$$\overline{\mathbf{G}} = \mathbf{R}_{inG}^{-1}\mathbf{C}(\mathbf{C}^H\mathbf{R}_{inG}^{-1}\mathbf{C})^{-1}\mathbf{g} \tag{62}$$

If the desired signal and interference are uncorrelated with the additive noise, $\overline{\mathbf{G}}$ can be written in the simple form (the proof is in Appendix C of Ref. [5])

$$\overline{\mathbf{G}} = \frac{g_1}{\nu}(2M|P_s|^2\overline{\mathbf{V}}_i^{(i)} - (P_s{}^{\Delta}P_i)_1 \boldsymbol{q}_s^H \boldsymbol{q}_i \overline{\mathbf{V}}_s^{(i)}) \tag{63}$$

where

$$\nu = (2M)^2|P_s|^2|P_i|^2 - |(P_i{}^{\Delta}P_s)_1|^2|\boldsymbol{q}_i^H\boldsymbol{q}_s|^2 = \mu - 2M\xi_i^{-1}\,|P_s|^2 \tag{64}$$

μ is given by (43). Moreover, the quaternion-valued optimal weight vector $\mathbf{G}_o$ may be given by

$$\mathbf{G}_o = J_1\,\overline{\mathbf{G}} + j\,J_2\,\overline{\mathbf{G}} \tag{65}$$

where $J_1 = [\mathbf{I}_{2M\times 2M},\ \mathbf{0}_{2M\times 2M}]$ and $J_2 = [\mathbf{0}_{2M\times 2M},\ \mathbf{I}_{2M\times 2M}]$ are two selection matrices.

By using the optimal weight vector $\mathbf{G}_o$, the complex output of second-stage beamformer can be given by

$$y_g(n) = (\mathbf{W}_o{}^{\Delta}\boldsymbol{v}_i)_1\ s_i(n) + (\mathbf{G}_o{}^{\Delta}\boldsymbol{n}^{(i)}(n))_1 \tag{66}$$

Thus, the complex output of QSWL GSC may be rewritten as

$$y_{GSC}(n) = y_w(n) - y_g(n) = s_s(n) + (\mathbf{W}_o{}^{\Delta}\ \boldsymbol{n}(n))_1 - (\mathbf{G}_o{}^{\Delta}\ \boldsymbol{n}^{(i)}(n))_1 \tag{67}$$

From above equation, we see that the interference component is completely cancelled in the output $y_{GSC}(n)$.

6.3. The performance analysis

Since the QSWL GSC can totally remove the interference, its output signal-to-interference ratio (SIR) tends to infinite. Thus, we focus our attention on the output signal-to-noise ratio (*SNR*) and array's gain. Let $\rho_n = \mathrm{E}\{|(\mathbf{W}_o^\Delta \boldsymbol{n}(n))_1 - (\mathbf{G}_o^\Delta \boldsymbol{n}^{(i)}(n))_1|^2\}$ is the power of output noise. From Eqs. (37) and (58), we have

$$(\mathbf{W}_o^\Delta \boldsymbol{n}(n))_1 - (\mathbf{G}_o^\Delta \boldsymbol{n}^{(i)}(n))_1 = (\mathbf{W}_1^H - \mathbf{G}_1{}^H)\boldsymbol{n}_1(n) + (\mathbf{W}_2^H + \mathbf{G}_2{}^H)\boldsymbol{n}_2(n) \tag{68}$$

Then, ρ_n can be written as

$$\rho_n = \sigma_n^2(\mathbf{W}_1^H - \mathbf{G}_1^H)(\mathbf{W}_1 - \mathbf{G}_1) + \sigma_n^2(\mathbf{W}_2^H + \mathbf{G}_2^H)(\mathbf{W}_2 + \mathbf{G}_2) \tag{69}$$

When the combined QPMC and MVDR are adopted in the second-stage beamformer, ρ_n can be written in the simple form (the proof is in Appendix D of Ref. [5])

$$\rho_n = \frac{\sigma_n^2}{\mu^2}(2M|P_s|^2\varepsilon^2 + |(P_i{}^\Delta P_s)_1|^2|\boldsymbol{q}_i^H\boldsymbol{q}_s|^2\lambda_q) \tag{70}$$

where $\lambda_q = \frac{\xi_i^{-2}|P_s|^2}{\kappa} - 2M|P_i|^2$, κ is given by Eq. (54). From Eq. (67), the expression of output SNR and array's gain A_q may be written as

$$SNR_o = \xi_s \frac{(2M|P_s|^2\varepsilon - |(P_i{}^\Delta P_s)_1|^2|\boldsymbol{q}_i^H\boldsymbol{q}_s|^2)^2}{(2M|P_s|^2\varepsilon^2 + |(P_i{}^\Delta P_s)_1|^2|\boldsymbol{q}_i^H\boldsymbol{q}_s|^2\lambda_q)} \tag{71}$$

$$A_q = \frac{(2M|P_s|^2\varepsilon - |(P_i{}^\Delta P_s)_1|^2|\boldsymbol{q}_i^H\boldsymbol{q}_s|^2)^2}{(2M|P_s|^2\varepsilon^2 + |(P_i{}^\Delta P_s)_1|^2|\boldsymbol{q}_i^H\boldsymbol{q}_s|^2\lambda_q)} \tag{72}$$

where ξ_s denotes the input signal-to-noise ratio (SNR) and ε is given by Eq. (43).

When the LCMV is adopted in the second-stage beamformer, ρ_n can be written in the simple form (the proof is in Appendix E of Ref. [5])

$$\rho_n = \frac{\sigma_n^2}{\mu^2}(2M|P_s|^2\varepsilon^2 + |(P_i{}^\Delta P_s)_1|^2|\boldsymbol{q}_i^H\boldsymbol{q}_s|^2\lambda_l) \tag{73}$$

where $\lambda_l = \frac{2M\xi_i^{-2}|P_s|^2}{\nu} - 2M|P_i|^2$, ν is given by Eq. (64). Then, the expression of output SNR and array's gain A_l may be written as

$$SNR_o = \xi_s \frac{(2M|P_s|^2\varepsilon - |(P_i{}^\Delta P_s)_1|^2|\boldsymbol{q}_i^H\boldsymbol{q}_s|^2)^2}{(2M|P_s|^2\varepsilon^2 + |(P_i{}^\Delta P_s)_1|^2|\boldsymbol{q}_i^H\boldsymbol{q}_s|^2\lambda_l)} \tag{74}$$

$$A_l = \frac{(2M|P_s|^2\varepsilon - |(P_i{}^\Delta P_s)_1|^2|\boldsymbol{q}_i^H\boldsymbol{q}_s|^2)^2}{(2M|P_s|^2\varepsilon^2 + |(P_i{}^\Delta P_s)_1|^2|\boldsymbol{q}_i^H\boldsymbol{q}_s|^2\lambda_l)} \tag{75}$$

From Eqs. (71), (72), (74) and (75), we can see that the output SNR and array's gain depend on not only separation between the DOA's of the desired signal and interference (i.e., $|\boldsymbol{q}_i^H\boldsymbol{q}_s|$), but also difference between the polarizations of the desired signal and interference (i.e., $|(P_i^\Delta P_s)_1|$). The dependencies of them on $|\boldsymbol{q}_i^H\boldsymbol{q}_s|$ and $|(P_i{}^\Delta P_s)_1|$ are shown in following consequences:

1. When $|\boldsymbol{q}_i^H \boldsymbol{q}_s| = 0$, the separation between the DOAs of the desired signal and interference reaches to maximum. In this case, $A_q = A_l = 2M|P_s|^2$. Further, $|\boldsymbol{q}_i^H \boldsymbol{q}_s|$ increases with a decrease of the DOA's separation. Thus, the array's gain A_q and A_l will reduce if $|P_s|^2$ is a constant. When $\boldsymbol{q}_i = \boldsymbol{q}_s$, $|\boldsymbol{q}_i^H \boldsymbol{q}_s| = 2M$. This implies that there is no separation between the DOAs of the desired signal and interference. In this case, the array's gain is given by

$$A_q = A_l = \frac{2M(|P_s|^2 \varepsilon - 2M|(P_i{}^{\Delta}P_s)_1|^2)^2}{|P_s|^2 \varepsilon^2 + 2M|(P_i{}^{\Delta}P_s)_1|^2 \lambda} \tag{76}$$

where

$$\lambda = \frac{\xi_i^{-2}|P_s|^2}{2M(|P_s|^2|P_i|^2 - |(P_i{}^{\Delta}P_s)_1|^2)} - 2M|P_i|^2 \tag{77}$$

Further, $P_i = P_s$ if $\gamma_s = \gamma_i$ and $\eta_s = \eta_i$. Thus, the array's gain $A_q = A_l = 0$ due to $\lambda = \infty$. This implies that the QSWL GSC fails.

2. When $|(P_i{}^{\Delta}P_s)_1| = 0$, $A_q = A_l = 2M|P_s|^2$. In the cases that $\theta_s = \theta_i \neq 0$ and $\varphi_s = \varphi_i \neq 0$ (i.e., $\boldsymbol{q}_i = \boldsymbol{q}_s$), we have $|(P_i{}^{\Delta}P_s)_1| = (\sin^2\theta_s \cos^2\varphi_s + \sin^2\varphi_s) \cos(\gamma_i - \gamma_s) \cos(\eta_i - \eta_s)$. If $\gamma_i - \gamma_s = \pm\pi/2$ or $\eta_i - \eta_s = \pm\pi/2$, then $|(P_i{}^{\Delta}P_s)_1| = 0$. This implies that even though there is no separation between the DOAs of the desired signal and interference, the array's gain can also reach to $2M|P_s|^2$ by using the orthogonality between the polarizations of the desired signal and interference. Further, the array's gain decreases with an increase of $|(P_i{}^{\Delta}P_s)_1|$ if $|P_s|^2$ is a constant. When $P_i = P_s$, $|(P_i{}^{\Delta}P_s)_1| = |P_s|^2$. This implies that there is no difference between the polarizations of the desired signal and interference. But, the array's gain is not equal to zero if $\boldsymbol{q}_i \neq \boldsymbol{q}_s$.

In addition, the output *SNR* and array's gain depend also on the input *INR* ξ_i, the array's element number $2M$, the interference response's power $|P_i|^2$ and the desired signal response's power $|P_s|^2$.

7. Monte Carlo simulations

In this section, we investigate the performance of the proposed beamformers by two experiments. More results of simulations were shown in Refs. [2, 4, 5].

7.1. Experiment 1: the performance of QMVDR beamformer

In practice, if there is a misalignment between the desired signal's DOA and the look direction, the SINR of the complex MVDR beamformer degrades in the case of a scalar vector array [13]. In this experiment, we investigate the robustness of the beamformer against the DOA mismatch.

http://dx.doi.org/10.5772/67859

We consider each two-component vector-sensor has two orthogonal magnetic loop co-aligned along the x-axis, and assume that $M = 1$. The result is the average of the output SINR obtained by *1000* Monte Carlo runs. To compare the performance, the complex 'long vector' MVDR (CLVMVDR) [13], QMVDR and its INC beamformers are included in simulation results. For the CLVMVDR and QMVDR beamformers, the SINR is in general expressed in the form of logarithm. In order to be identical with the SINR of CLVMVDR and QMVDR beamformers, we define $SINR = 10\log\frac{\sigma_s^2}{\sigma_i^2+\sigma_n^2}$ in this experiment and assume that input $SINR = 0$ dB.

Figure 5 displays the output SINR as the function of the DOA error of the desired signal where the DOA error is between $-8°$ and $8°$. In the case of $\theta_s = 0°, \theta_i = 80°, \varphi_s = \varphi_i = 0°$; $\gamma_s = \gamma_i = 60°, \eta_s = \eta_i = 30°$, (i.e., the polarization of the desired signal and interference is identical, but the DOA is not identical), simulation results for two different sample sizes $N = 20$ and $N = 500$ are given in **Figure 5(a)** and **(b)**, respectively. From **Figure 5**, it is seen that the output *SINR* behaviour is different with different values of sample size N. In case of small N, as shown in **Figure 5(a)**, the INC and the QMVDR have a better robustness than the CLVMVDR. But the output *SINR* of the QMVDR and CLVMVDR is more than that of the INC. In case of large N, as shown in **Figure 5(b)**, the robustness against the DOA mismatch is almost identical for three beamformers. But the INC has the largest output *SINR* in three beamformers.

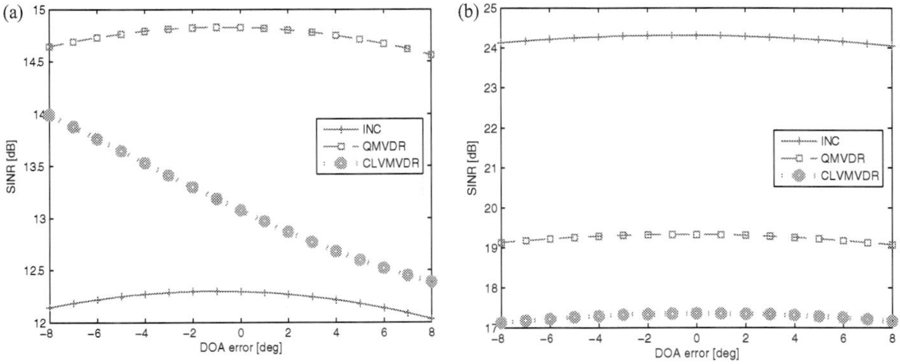

Figure 5. Output *SINR* against the DOA error.

7.2. Experiment 2: the performance of QSWL beamformer

In the second experiment, we illustrate the performance of the proposed QSWL GSC in the presence of a single interference. We assume $M = 6$, $\varphi_s = \varphi_i = 60°$; $\gamma_s = \gamma_i = 30°, \eta_s = \eta_i = 30°$, and that the covariance matrix $\mathbf{R}_x$, instead of $\mathbf{R}_{in}$, is available. **Figure 6** displays the power patterns for three values of $|\Delta\theta|$: $60°$, $20°$ and $10°$, where $\theta_s = |\Delta\theta|, \theta_i = 0°$. From **Figure 6**, it is seen that three beamformers steer almost a zero towards the interference's DOA (located at 0°) in all cases. When $|\Delta\theta|$ decreases, the main-lobe of the QSWL GSC points almost to the source location, but the main-lobe of the complex 'long vector' LCMV (CLCMV) is away from

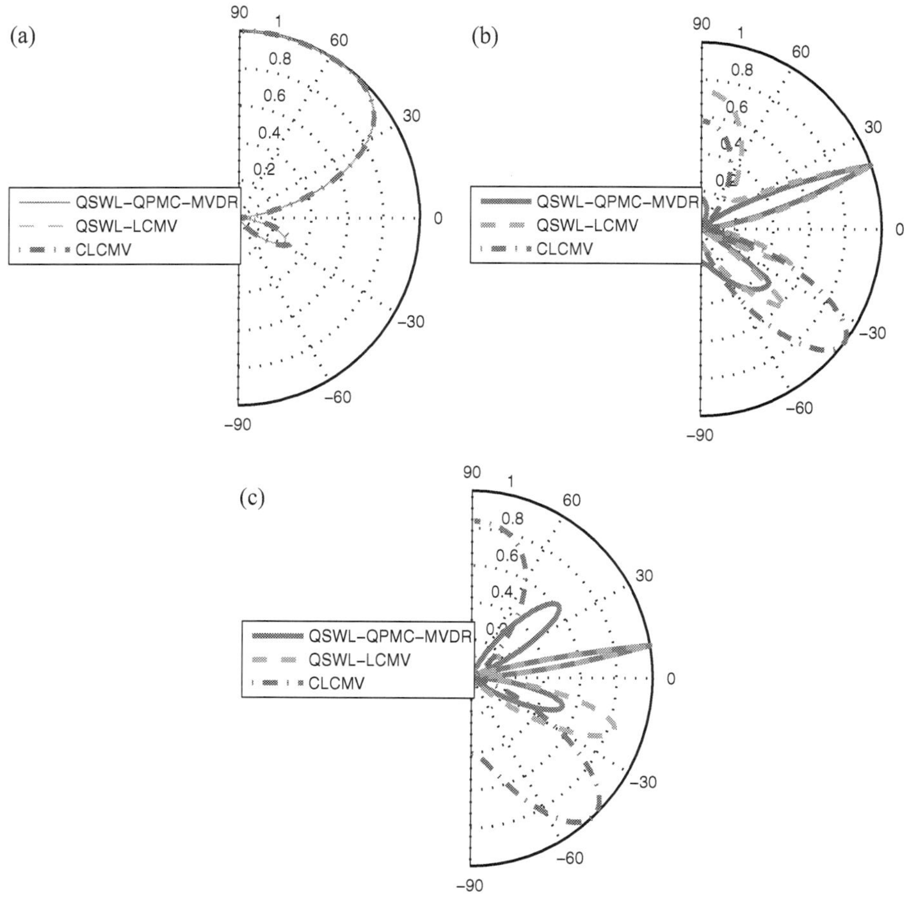

Figure 6. The power patterns at $\theta_s = |\Delta\theta|, \theta_i = 0°$. (a) $|\Delta\theta| = 60°$, (b) $|\Delta\theta| = 20°$, (c) $|\Delta\theta| = 10°$.

the source location. This implies that the QSWL GSC outperforms obviously CLCMV as the desired signal moves towards the interference. In addition, the side-lobes are amplified with a decrease of $|\Delta\theta|$. These side-lobes lead the beamformer to capture the white noise, which spans the whole space, so that the performance of beamformer degrades.

8. Conclusion

The problem of beamformer based on quaternion processes is considered in this chapter. The quaternion beamformers has more information than the complex 'long vector' beamformer.

http://dx.doi.org/10.5772/67859

The increase of information results in the improvement of the beamformer's performance. Analyses in theory and simulation results verify the advantages of quaternion beamformers.

Acknowledgements

This work was supported by the National Natural Science Foundation of China under Grants 61571462 and 60872088.

Author details

Jian-wu Tao[1,2] and Wen-xiu Chang[1,2]*

*Address all correspondence to: wenxiu_chang@gmail.com

1 Aviation University, Changchun, China

2 Jilin University, Changchun, China

References

[1] Tao, J.W., and Chang, W.X., "Quaternion MMSE Algorithm and Its Application In Beamforming", *Acta Aeronautica et Astronautica Sinica*, 2011, Vol. 32, no. 4, pp. 729–738.

[2] Tao, J.W., and Chang, W.X., "The MVDR Beamformer Based on Hypercomplex Processes", *Proc. 2012 IEEE International of Conference on Computer Science and Electronic Engineering*, Mar. 2012, pp. 273–277.

[3] Tao, J.W., and Chang W.X., "A Novel Combined Beamformer Based on Hypercomplex Processes", *IEEE Transactions on Aerospace and Electronic Systems*, 2013, Vol. 49, no. 2, pp. 1276–1288.

[4] Tao, J.W., "Performance Analysis for Interference and Noise Canceller based on Hypercomplex and Spatio-temporal-polarisation Processes", *IET Radar, Sonar and Navigation*, 2013, Vol. 7, no. 3, pp. 277–286.

[5] J.W. Tao, and W.X. Chang, "The Generalized Sidelobe Ganceller Based on Quaternion Widely Linear Processing", *The Scientific World Journal*, 2014, Vol. 2014, Article ID 942923, 12 pages.

[6] Mir, H.S., and Sahr, J.D., "Passive Direction Finding Using Airborne Vector Sensors in the Presence of Manifold Perturbations", *IEEE Transactions on Signal Processing*, 2007, Vol. 55, no. 1, pp. 156–164.

[7] Xiao, J.J., and Nehorai, A., "Optimal Polarized Beampattern Synthesis using a Vector Antenna Array", *IEEE Transactions on Signal Processing*, 2009, Vol. 57, no. 2, pp. 576–587.

[8] Nehorai A., Ho K.C., and Tan B.T.G., "Minimum-noise-variance Beamformer with an Electromagnetic Vector Sensor", *IEEE Transactions on Signal Processing*, 1999, Vol. 47, no. 3, pp. 601–618.

[9] Wong, K.T., "Blind Beamforming/Geolocation for Wideband-FFHs with Unknown Hop-Sequences", *IEEE Transactions on Aerospace and Electronic Systems*, 2001, Vol. 37, no. 1, pp. 65–76.

[10] Ward, J.P., *Quaternions and Cayley Numbers: Algebra and Applications*. Boston, MA: Kluwer, 1997.

[11] Mandic, D.P., Jahanchahi, C., and Took, C.C., "A Quaternion Gradient Operator and Its Applications," *IEEE Signal Processing letters*, 2011, Vol. 18, no. 1, pp. 47–50.

[12] Vía, J., Ramírez, D., and Santamaría, I., "Properness and Widely Linear Processing of Quaternion Random Vectors," *IEEE Transactions on Information Theory*, 2010, Vol. 56, no. 7, pp. 3502–351.

[13] Van Trees, H., *Optimum Array Processing (Detection Estimation, and Modulation Theory), Part IV.* New York: Wiley-Interscience, 2002.